Alabama
the Heart of Dixie

★ Montgomery

camellia

yellowhammer

Since becoming the 22nd state in 1819, Alabama has experienced great change. Once a one-crop (cotton) state, its farm income today consists primarily of such things as broiler chickens, beef cattle, soybeans, corn, peanuts and strawberries.

Its manufacturing industry, no longer dependent solely on steel, now includes textiles, aluminum, paper and chemical products.

Race relations have long been a problem, but blacks are beginning to play a greater role in politics. This change was begun by the famous Martin Luther King, Jr.

MORE TO LEARN

B	C	A	T	T	L	E	C	A
R	H	S	E	W	H	E	P	P
O	I	T	X	V	C	D	A	E
I	C	O	T	T	O	N	P	A
L	K	C	I	N	R	L	E	N
E	E	O	L	X	N	L	R	U
R	N	R	E	W	H	E	A	T
S	S	O	Y	B	E	A	N	S
C	H	E	M	I	C	A	L	M

Circle the various crops, products, etc. mentioned above in the **WORD SEARCH**. There are ten, going → or ↓ . Anytime a word is circled that has a box around a letter, write that letter below.

___ ___ ___ ___ ___ ___ ___ ___ ___ ___

Now, unscramble the letters to discover the last name of the four-time governor who tried to stop school integration.

___ ___ ___ ___ ___ ___ ___

CAN YOU FIND...

What baseball player from Alabama holds the home run record? _____

1

Alaska
the Last Frontier

Name _____

Alaska, the 49th state, ranks first in size and last in population in the nation! Juneau, its capital, is the nation's largest city in area— over 3,000 square miles — but has only 19,000 people.

Alaska touches no other state. Instead, it is bordered on the north by the Arctic Ocean, on the south by the Pacific Ocean, on the east by Canada and on the west by the Bering Sea. At one point it is only 51 miles across the Bering Strait to Russia.

The vast untold natural resources and incredible beauty of this "Land of the Midnight Sun" make it truly America's last frontier.

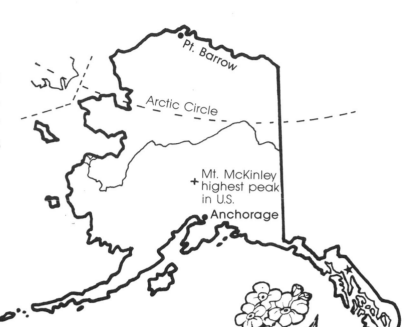

forget-me-not

Label the bodies of water, bordering country, nearby country and state capital on the map.

MORE TO LEARN

Mark each statement **TRUE** OR **FALSE**. On another sheet of paper, rewrite each FALSE statement so it is TRUE.

willow ptarmigan

_____ Alaska touches one other state.

_____ The state ranks first in size in the nation.

_____ Anchorage is the largest city and the state capital.

_____ Juneau has the largest population of any city in the nation.

_____ Alaska ranks last in population of the fifty states.

_____ Mt. McKinley is the highest peak in the world.

_____ Canada borders Alaska on the east.

_____ The state is bordered by two oceans.

CAN YOU FIND . . . What was the 1867 purchase of Alaska for $7,200,000 called by Americans? _____

Arizona
the Grand Canyon State

cactus wren

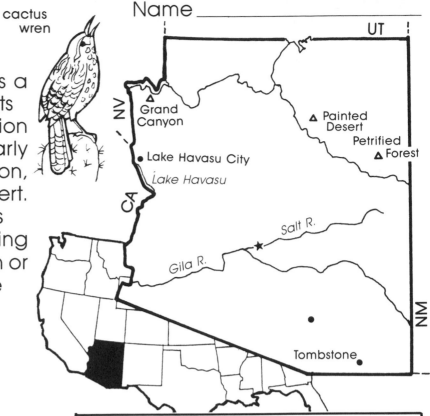

Arizona, the 48th state, has a history greatly influenced by its need for water. The first irrigation of its desert was by Indians nearly 800 years ago. Without irrigation, half the state would be a desert.

The warm, dry climate has made Arizona a rapidly-growing state. Most of the people live in or near Tucson and Phoenix, the capital and largest city.

A history of Mexican and Indian influence is quite apparent in Arizona's culture.

saguaro blossom

On the map, label the following: state capital, second largest city, bordering country and river running through the Grand Canyon.

MORE TO LEARN

Write the number of the phrase in Column **B** that describes who or what each is in Column **A**. Think logically!

A	B
Sandra Day O'Connor ___	1. Indian chief who kept fighting
Tombstone ___	2. Located at·Lake Havasu City (but from England)
Geronimo ___	3. River running through the Grand Canyon
Barry Goldwater ___	4. Where Wyatt Earp won fame
Colorado ___	5. First female U.S. Supreme Court Justice
Navajo ___	6. Largest cactus in U.S.
Saguaro ___	7. Largest Indian tribe in Arizona
The London Bridge ___	8. Father of Arizona Republican Party

CAN YOU FIND... What are the names of two national monuments in Arizona?

Arkansas
the Land of Opportunity

Arkansas, the 25th state, is a land rich in natural resources, both above and beneath the soil. Its lowlands have rich soil for farming. Its plateau area is suitable for raising hogs, cattle and broiler chickens (the country's leader). One quarter of the state is covered with rich timber.

Beneath the soil is found nearly all the nation's bauxite, the only diamond mine, and deposits of petroleum and natural gas.

The state's natural beauty attracts millions of tourists each year to this "Land of Opportunity".

apple blossom

MORE TO LEARN

Use the pictures on the map and the clues below to identify the towns, cities, bordering states and points of interest. Write the numbers next to the △'s on the map.

1. Little Rock, the state capital
2. Pea Ridge, site of a Civil War battle
3. Murfreesboro, town near only diamond mine in the U.S.
4. Hot Springs National Park
5. Town of Hot Springs
6. El Dorado, site of first oil discovery in the state
7. Springdale, in center of broiler chicken area
8. Bauxite, site of first aluminum ore mine
9. Missouri - north
10. Louisiana - south
11. Oklahoma - west
12. Tennessee - northeast
13. Texas - southwest
14. Mississippi - southeast

mockingbird

CAN YOU FIND . . . Who was the Arkansas woman that was the first female elected to the U.S. Senate? _____

California
the Golden State

California, the 31st state, is almost beyond comprehension! In America, this state has the largest population, the most goods produced, the highest agriculture output, the tallest and oldest living things, the largest city (Los Angeles) — the list is endless.

Also a land of great natural beauty, California's Yosemite National Park contains Ribbon Falls, the highest waterfall in North America, and Mount Whitney, the highest point in the U.S. south of Alaska. Yet its Death Valley is the lowest point in North America! Its 840 miles of coastline varies from steep cliffs to sandy beaches with two great natural harbors — San Francisco and San Diego.

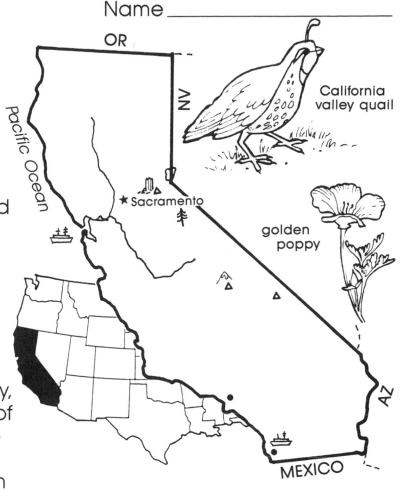

California valley quail

golden poppy

Label the cities and points of interest on the map.

MORE TO LEARN

R	S	A	N	T	S	N	E	V	A	D	A
I	Y	O	S	E	M	I	T	E	T	W	R
B	W	H	I	T	N	E	Y	S	D	V	I
B	D	E	A	T	H	V	R	A	I	R	Z
O	V	A	L	L	E	Y	P	N	E	B	O
N	X	M	N	S	L	O	R	E	G	O	N
S	A	C	R	A	M	E	N	T	O	N	A
B	F	R	A	N	C	I	S	C	O	V	P

Locate the following in the **WORD SEARCH:**

- The bordering states
- The two natural harbors
- Lowest point in North America
- Highest waterfall in North America
- State capital
- Highest mountain in U.S. south of Alaska
- A national park

CAN YOU FIND . . . Who was the famous English sea captain who sailed around the world claiming what is now California for England in 1579? _____

Colorado
the Centennial State

Colorado, the 38th state, is the highest state in the nation. It has the highest road in the U.S. and the world's highest tunnel for vehicles.

The Rocky Mountains, cause of the transportation problems, also attract tourism, provide vast mineral resources and are the source of water for the plains area.

A fast-growing state, Colorado's need for water storage has been a severe and ongoing problem. Water is the lifeblood of the people.

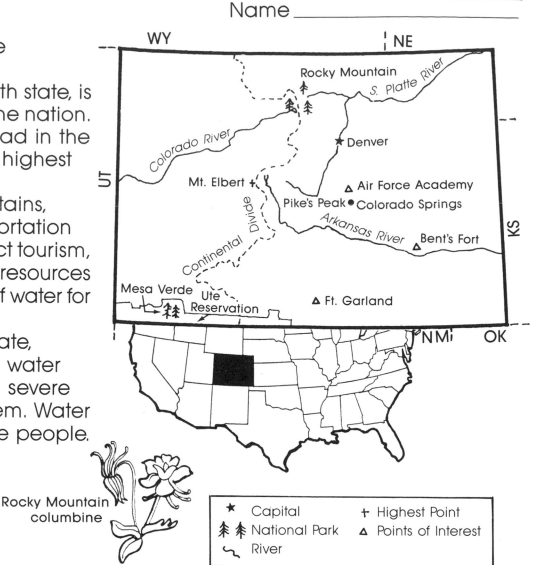

lark bunting

Rocky Mountain columbine

★ Capital   ✝ Highest Point
🌲🌲 National Park   △ Points of Interest
〜 River

MORE TO LEARN

Which rivers flow east? _____ west? _____
Which state borders on the east? ___ west? ___ northeast? ___
southeast? ___ northwest? ___ southwest? ___
Colorado's mile-high capital is _____.
The oldest residents of Colorado are the _____ Indians who live on a reservation in the southwest corner of the state.
The highest mountain is _____. The second highest is _____ .
The U.S. Air Force Academy is near _____ .
Kit Carson long ago commanded the historic posts _____ and
_____ .
The two national parks are _____ and _____ .

CAN YOU FIND . . .   Why is Colorado known as "the Centennial State"?
_____

6

Connecticut
the Constitution State

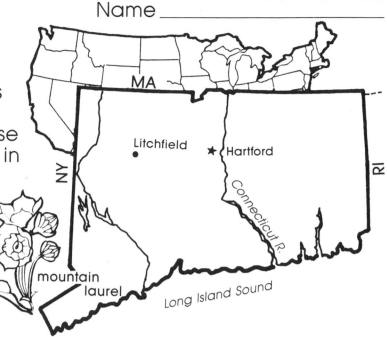

Litchfield • ★ Hartford

MA

NY

Connecticut R.

RI

mountain laurel

Long Island Sound

Connecticut, the 5th state, has a history of innovation. This trait accounts for its nickname because the Fundamental Orders, drafted in 1638, was the first constitution in the New World.

Industrialization got an early start with mass production and steel manufacturing beginning there. The first insurance for accidents and automobiles was written in Hartford, now known as the "Insurance City". And, Connecticut established the first American public school!

MORE TO LEARN

Using the **WORD BANK**, fill in the blanks to learn more about Connecticut events and people. The circled letters, when unscrambled, spell the last name of the man who first vulcanized rubber, Charles _____.

WORD BANK			
Oliver Cromwell	Nautilus	Litchfield	Mark Twain
Katherine Hepburn	Nathan Hale	Eli Whitney	hamburger

_ _ _ _ _ _ _ _ _ _ _ _ _, hung as spy during the Revolutionary War

○_ _ _ _ _ _ _○_ _ _ _ _ _, the first American warship

_○_ _ _ _ _ _ _, first automic submarine

robin

_ _ _ _○_ _ _ _ _ _ _ _ _ _ _ _ _, famous actress

_ _○_ _ _ _ _ _ _ _, wrote Huckleberry Finn there.

_ _ _ _ _ _ _ _ _ _○, site of America's first law school

_ _ _ _ _ _ _ _ _ _○ made the state the birthplace of mass production.

_ _ _ _ _ _○_ _, a favorite American food, first eaten there

CAN YOU FIND . . .

What famous circus owner is from Connecticut?

_____

Delaware
the First State

Name _____

peach blossom

Atlantic Ocean

The only state to belong to Sweden and Holland, England won the area that is known now as Delaware from the Dutch in 1664. About 100 years later, the American flag was first displayed in battle at the Revolutionary Battle of Cooch's Bridge. Named for Lord De La Warr, Delaware was the first state to ratify the new constitution in 1787.

Wilmington, the largest city, is the home of E. I. du Pont de Nemours and Co., the world's largest manufacturer of chemicals. The DuPont family has had great economic and political influence on the state for many years.

Write the names of the cities and points of interest labeled on the map.

_____   _____

_____

_____

blue hen chicken

MORE TO LEARN

D _____, the state capital

E _____, the world's largest chemical manufacturer

L _____, for whom state was named

A _____ Ocean, bordering on the east

W _____ Museum, location of Art Conservation Project

A merican flag first displayed in battle at _____, September 3, 1777

R _____ the new constitution first!

E _____ won the area from the Dutch in 1664.

CAN YOU FIND . . . What county in Delaware leads the nation in production of broiler chickens? _____

Florida
the Sunshine State

AL
GA
Gulf of Mexico
★ Tallahassee
Jacksonville
St. Augustine ●
mockingbird
Cape Canaveral
St. Petersburg
● Tampa
Lake Okeechobee
Atlantic Ocean
Ft. Lauderdale
● Miami
Everglades National Park

The area that is now Florida, becoming the 27th state in 1845, has an early history. Ponce de Leon first explored it in 1513. Fifty years later the French built Fort Caroline. But in 1565, the Spanish defeated the French and founded St. Augustine, the oldest city in the U.S. Spain ceded Florida to England who in turn traded it back in 1783 for the Bahamas!

But Florida is also the site of much modern history, with Cape Canaveral launching the first American in space in 1961, and the first American on the moon in 1969.

orange blossom

MORE TO LEARN

Number the historic events in the order they occurred.

___ Flo_rida became a state.

___ Spain ceded Fl©rida to England.

___ Firs_t Am©rican in space.

___ French bu_ilt Fort Caro①ine.

___ England traded Flor_ida to Spain.

___ Ponce de Leon explor_ed the a®ea.

___ Spanish de①eated the French.

___ First American ⓦalked on the _moon.

___ Spanis_h founded St. Augu⑤tine.

Write the underlined letters.

___ ___ ___ ___ ___ ___ ___

Unscramble them to discover Florida's leading industry.

___ ___ ___ ___ ___ ___ ___

Write the circled letters.

___ ___ ___ ___ ___ ___

Unscramble them to discover the meaning of the name, Florida. "full of

___ ___ ___ ___ ___ ___ ___ "

CAN YOU FIND . . . What is the name of the world's first oceanarium?

_____

9

Georgia the Empire State

Georgia, the 4th state, has come a long way in handling its long term racial problems. Two great leaders contributed much to integration — the late Martin Luther King, Jr. and Andrew Young, Ambassador to the United Nations in 1977.

Being an agrarian state, Georgia is now dominated by service industries (wholesale and retail trade). Farming is still important as it is a leading producer of tobacco, peaches and peanuts.

The state's most famous peanut farmer is Jimmy Carter — former President of the United States.

TN   NC
Blue Ridge Mountains
★ Atlanta
AL
FL
Atlantic Ocean
brown thrasher
Cherokee rose

MORE TO LEARN

Use the code to learn some interesting people, places and events.

1 - F	2 - M	3 - P	4 - B	5 - Z	6 - L	7 - A	8 - T	9 - V
10 - C	11 - N	12 - W	13 - S	14 - D	15 - E	16 - R	17 - G	18 - J
19 - H	20 - K	21 - Q	22 - I	23 - U	24 - Y	25 - X	26 - O	

Huge bird sanctuary
_ _ _ _ _ _ _ _ _ _   _ _ _ _ _
26 20 15 1 15 11 26 20 15 15   13 12 7 2 3

Route of Cherokee nation to Oklahoma
_ _ _ _ _   _ _   _ _ _ _ _
8 16 7 22 6   26 1   8 15 7 16 13

U. S. President who died at Warm Springs
_ _ _ _ _ _ _   _ .   _ _ _ _ _ _ _ _ _
1 16 7 11 20 6 22 11   14 .   16 26 26 13 15 9 15 6 8

Union General who captured Atlanta
_ _ _ _ _ _ _   _ _ _ _ _ _ _
12 22 6 6 22 7 2   13 19 15 16 2 7 11

First female U. S. Senator
_ _ _ _ _ _ _   _ .   _ _ _ _ _ _
16 15 4 15 10 10 7   6 .   1 15 6 8 26 11

Georgia woman who wrote *Gone With the Wind*
_ _ _ _ _ _ _ _   _ _ _ _ _ _ _ _
2 7 16 17 7 16 15 8   2 22 8 10 19 15 6 6

CAN YOU FIND . . .

Who invented the cotton gin near Savannah in 1793?
_____

Hawaii the Aloha State

Name _____

Kauai

Pacific Ocean

hibiscus

Honolulu
Oahu

Molokai

Lanai

Maui

Kahoolawe

nene
(Hawaiian goose)

Hawaii, the 50th state, is a group of 132 islands formed from volcanic mountains. Diamond Head is one of its most famous extinct volcanoes.

An important naval base, Pearl Harbor, was the first part of the U.S. territory to be attacked in World War II.

The state's economy is heavily dependent on the service industries (nearly 90%), with tourism being the most important. Hawaii's culture is unique, with its colorful dress, famous luaus, hula dance and use of many polynesian terms such as "malihini" (newcomer). Its alphabet consists of only twelve letters with two consonants never coming together in a word.

Hawaii

List the seven inhabited islands in order of size. _____

_____

MORE TO LEARN

Mark each statement **(T)** TRUE or **(F)** FALSE. On another sheet of paper, rewrite each FALSE statement so it is TRUE.

___ Hawaii is a group of islands formed from active volcanic mountains.
___ This is the only state that was an independent monarchy.
___ Hawaii was the only part of U.S. territory to be attacked during World War II.
___ Diamond Head is an active volcano.
___ Hawaii was the last state admitted to the Union.
___ The most famous craft is the making of pottery products.
___ Service industries account for nearly 90% of the state's gross.

CAN YOU FIND . . . What was the name given to the islands in 1778 by the English sea captain James Cook? _____

Idaho the Gem State

Idaho, the 43rd state, is famous across the country for its "Idaho potatoes". The state also leads the nation in production of silver and lead. Rich in natural beauty and natural resources (especially water), tourism and manufacturing are growing rapidly.

As Idaho continues to become more urban, the people will be challenged to preserve their beautiful wilderness land full of such wonders as Hell's Canyon, deeper than the Grand Canyon, and Shoshone Falls, higher than Niagara!

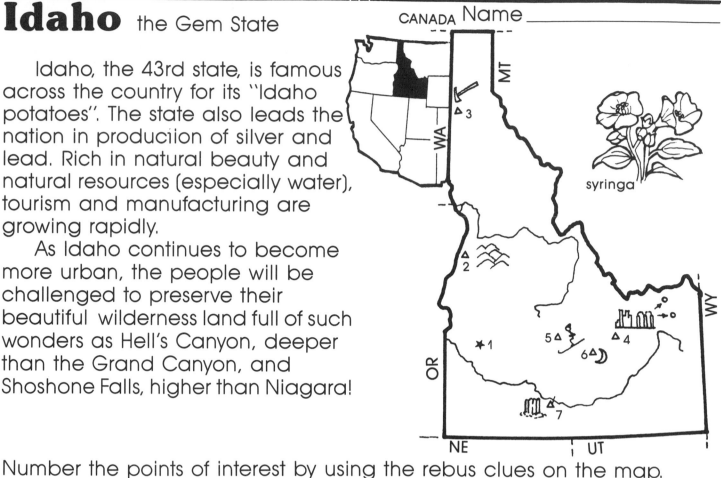

syringa

Number the points of interest by using the rebus clues on the map.

___ Boise
___ Shoshone Falls
___ Arco Atomic Power Station
___ Crater of the Moon National Monument

___ Hell's Canyon
___ Sun Valley
___ Nation's largest silver mine

MORE TO LEARN

Make each statement **TRUE** or **FALSE**. On another sheet of paper, rewrite each FALSE statement so it is TRUE.

_____ Idaho leads the nation in production of gold and silver.

_____ The state is becoming more rural.

_____ Potatoes are one of its most famous products.

_____ Abundant water is an important natural resource.

_____ The nation's largest silver mine is in Idaho.

_____ The Grand Canyon is the world's deepest canyon.

_____ Boise is the capital of Idaho.

_____ Sun Valley is a famous beach.

mountain bluebird

CAN YOU FIND...

What former governor was murdered during a mining dispute? _____

Illinois

the Prairie State

Name_____

Illinois, the 21st state, is on one hand agrarian, leading the country in soybean and corn production. In contrast is Chicago, with the world's greatest railroad center, busiest airport and tallest building. There has always been a battle between the rural areas and Chicago for political control.

Illinois products are carried to many parts of the world, either down the Chicago River to Lake Michigan and the St. Lawrence Seaway, or the Illinois River which links with the Mississippi River and Gulf of Mexico.

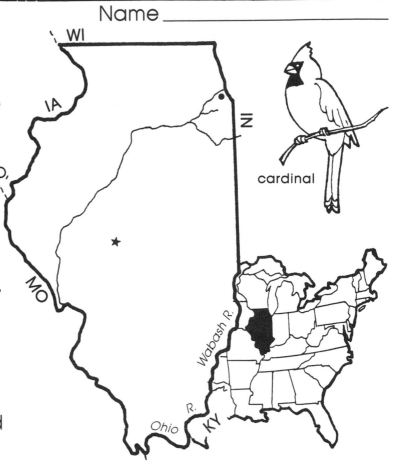

cardinal

violet

Label Springfield (the state capital), the largest city, three rivers and the Great Lake.

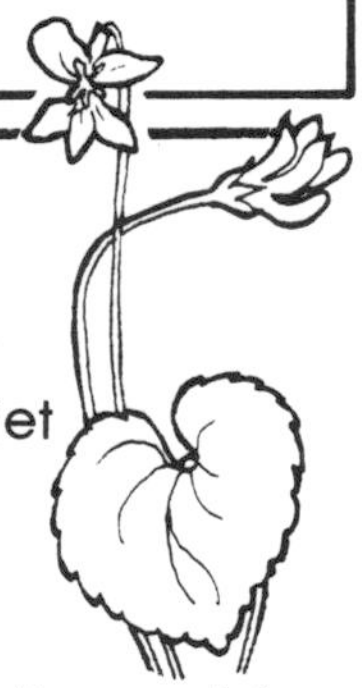

MORE TO LEARN

Use the rhyming clues to fill in the blanks.

Th(e) world's busiest (a)irport, O'H_____ (rhymes with mare)

Chicago's gift to black l(e)adership, Jesse J_____ (rhymes with hack and ton)

Famou(s) railroad industrialist, George P_____ (rhymes with bull and tan)

(W)orld's largest mapmaker R_____ McN_____ (rhymes with band and tally)

Famous U(n)ion general, Ulysses S. Gr_____ (rhymes with plant)

Most i(m)portant natural resource, s_____ (rhymes with toil)

Nation's (l)argest deposits of bituminous c_____ (rhymes with foal)

Write the circled letters. _____ Unscramble them to find the name of Lincoln's boyhood town which is now restored. _____

CAN YOU FIND . . . What President was born in Tampico, Illinois?

_____

Indiana
the Hoosier State

peony

Name _____

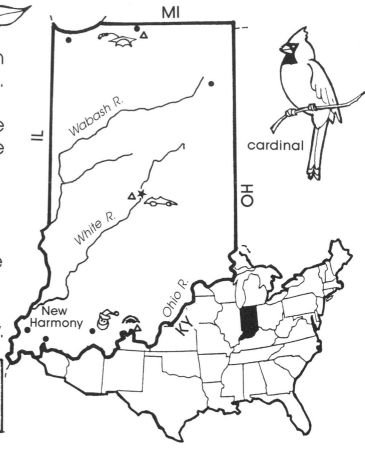

Indiana, the 19th state, ranks 38th in the U.S. in area, but 12th in population. Its largest city and state capital is Indianapolis. The other large cities are Ft. Wayne in the northeast, Gary in the northwest and Evansville in the southwest. Heavily industrialized, manufactured products account for 84% of the total value of goods.

In South Bend is the famous Notre Dame University. In the south is huge Wyandotte Cave, and outside Indianapolis is the Indy 500 Raceway.

Label the cities and points of interest marked on the map.

MORE TO LEARN

Choose words from the **WORD BANK** to fill in the boxes.

NEW HARMONY	CORN	MEMORIAL
STUDEBAKER	SOIL	BEN-HUR
HARRISON	SANTA CLAUS	

Town famous for its Christmas postmark
Famous book by Lew Wallace
Most valuable farm product
Brothers who made cars in 1902
Famous automobile race held on _____ Day
One of Indiana's greatest natural resources
Experimental community begun in 1825
Last name of two related U.S. Presidents

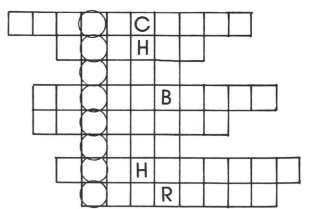

The circled letters spell the name of the Shawnee Indian chief defeated at the Battle of Tippecanoe. _____

CAN YOU FIND... What city was begun by the United States Steel Corporation in 1906? _____

Iowa
the Hawkeye State

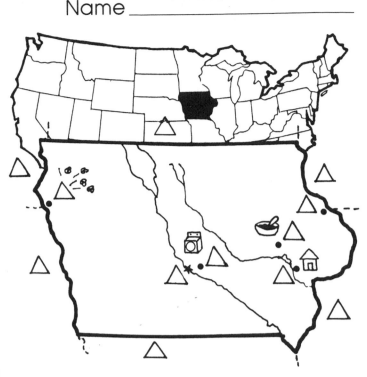

Iowa, the 29th state, is a farm state known as "the land where the tall corn grows." Its fertile soil makes the state a leading producer of corn, soybeans, beef cattle, hogs and dairy products.

The state's most important industries are related to agriculture, such as the manufacture of farm machinery and the processing of food products.

Iowa leads the nation in literacy — nearly everyone can read and write. Its school system has produced such famous people as President Herbert Hoover, Vice President (under President F. D. Roosevelt) Henry Wallace and artist Grant Wood.

MORE TO LEARN

Use the pictures on the map and the clues below to identify the towns, cities, bordering states and points of interest. Write the numbers in the △ 's on the map.

1. Des Moines, largest city and state capital
2. Sioux City, nation's largest popcorn-processing plant
3. West Branch, site of Herbert Hoover's birthplace
4. Cedar Rapids, nation's largest cereal mill
5. Newton, nation's largest home-laundry appliance factory
6. Dubuque, named after Iowa's first settler
7. Minnesota — north
8. Nebraska — southwest
9. Illinois — southeast
10. Missouri — south
11. Wisconsin — northeast
12. South Dakota — northwest

eastern goldfinch

wild rose

CAN YOU FIND . . .   Why is Iowa called the Hawkeye State? _____

Kansas the Sunflower State

Name _____

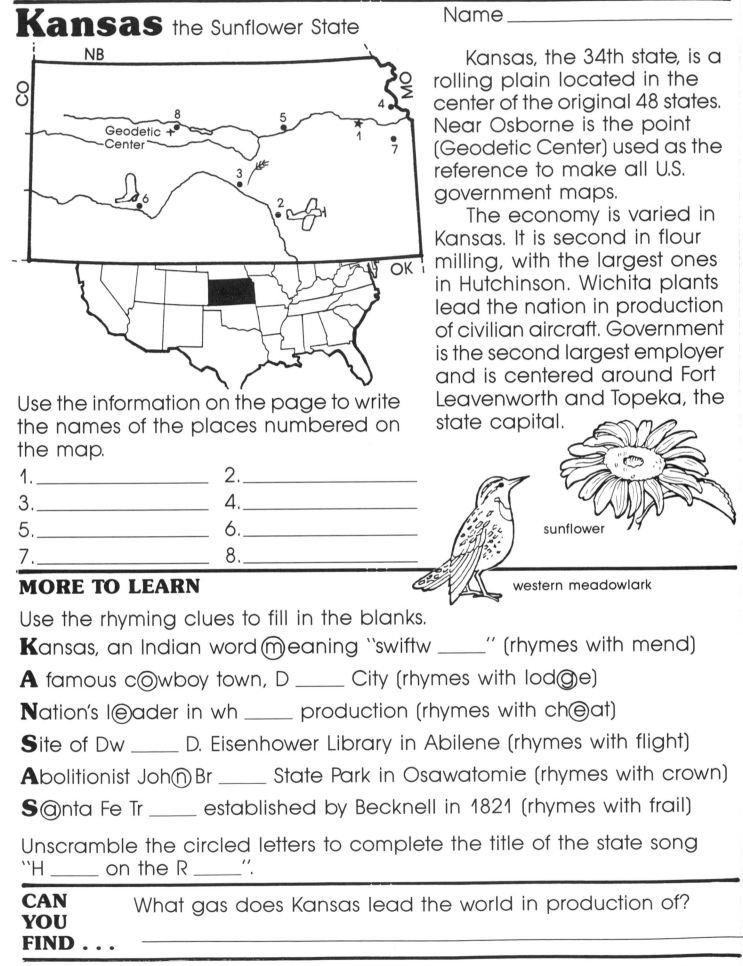

Kansas, the 34th state, is a rolling plain located in the center of the original 48 states. Near Osborne is the point (Geodetic Center) used as the reference to make all U.S. government maps.

The economy is varied in Kansas. It is second in flour milling, with the largest ones in Hutchinson. Wichita plants lead the nation in production of civilian aircraft. Government is the second largest employer and is centered around Fort Leavenworth and Topeka, the state capital.

Use the information on the page to write the names of the places numbered on the map.

1. _____ 2. _____
3. _____ 4. _____
5. _____ 6. _____
7. _____ 8. _____

sunflower

western meadowlark

MORE TO LEARN

Use the rhyming clues to fill in the blanks.

Kansas, an Indian word ⓜeaning "swiftw _____" (rhymes with mend)

A famous cⓞwboy town, D _____ City (rhymes with lodⓖe)

Nation's lⓔader in wh _____ production (rhymes with chⓔat)

Site of Dw _____ D. Eisenhower Library in Abilene (rhymes with flight)

Abolitionist Johⓝ Br _____ State Park in Osawatomie (rhymes with crown)

Sⓐnta Fe Tr _____ established by Becknell in 1821 (rhymes with frail)

Unscramble the circled letters to complete the title of the state song "H _____ on the R _____".

CAN YOU FIND . . . What gas does Kansas lead the world in production of?

_____

16   © 1992 Instructional Fair, Inc.

Kentucky
the Bluegrass State

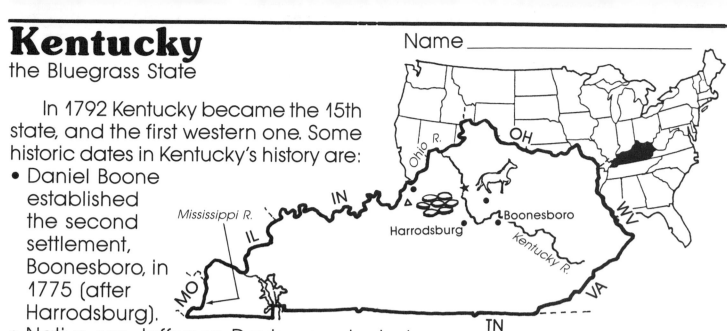

In 1792 Kentucky became the 15th state, and the first western one. Some historic dates in Kentucky's history are:

- Daniel Boone established the second settlement, Boonesboro, in 1775 (after Harrodsburg).
- Native son Jefferson Davis was elected President of the Confederacy in 1861.
- The year before, native son Abraham Lincoln was elected U.S. President.
- The first Kentucky Derby was held in 1875, about 50 years after William McGuffey produced his first *McGuffey Reader*.
- The nation's gold vault was established at Fort Knox in 1936.

> On the map label Frankfort, the capital, Fort Knox, Louisville and Lexington, the center of horse country.

MORE TO LEARN

Number the historic events listed below in the order they occurred.
___ Lincoln elected U.S. President
___ First Kentucky Derby held.
___ Boonesboro settled.
___ Nation's gold vault established.
___ Harrodsburg settled.
___ First *McGuffey Reader* produced.
___ Kentucky gained statehood.
___ Davis elected Confederate President.

cardinal

goldenrod

Write the bordering states in alphabetical order. _____

_____

CAN YOU FIND . . . What is located in Kentucky that is known as one of the Seven Natural Wonders of the modern world? _____

Louisiana

the Pelican State

Name _____

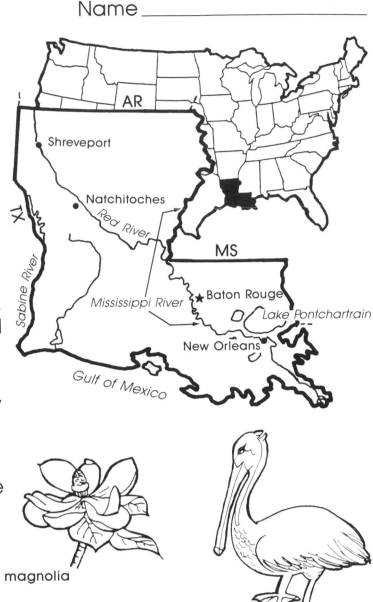

magnolia

eastern brown pelican

In 1812 Louisiana became the 18th state — the first one out of the original Louisiana Territory. Some historic dates in Louisiana's history are:

- LaSalle claimed the area for France in 1682 and named it.
- In 1718 de Bienville founded New Orleans.
- In 1760 the first Acadians came, two years before France ceded the land to Spain.
- In 1803 the U.S. bought the Louisiana Territory from France, three years after Spain had ceded it back to them.
- Two years before becoming a state, the eastern lands became the Republic of West Florida.
- In 1901 oil was discovered, forty years after Louisiana joined the Confederacy.

MORE TO LEARN

Number the historic events in the order they occurred.

___ New Orleans founded.

___ Louisiana became a state.

___ U. S. bought the Territory.

___ Republic of West Florida established.

___ Area named Louisiana.

___ Oil was discovered.

___ Louisiana joined the Confederacy.

___ Acadians first came.

___ France ceded area to Spain.

___ Spain ceded area to France.

CAN YOU FIND . . .   What is a county in Louisiana called? _____

Our 50 States IF8749   18   © 1992 Instructional Fair, Inc.

Maine

the Pine Tree State

Maine, the 23rd state, is the first to greet the sun each day in the United States. It is a land of great natural beauty with its rugged coastline, sparkling lakes, rushing rivers and magnificent mountains. This beauty attracts tourists year round.

Ninety percent of the land is covered by woods which provides the product for its huge lumbering industry. Maine leads the nation with its lobster catch — a great delicacy in this country.

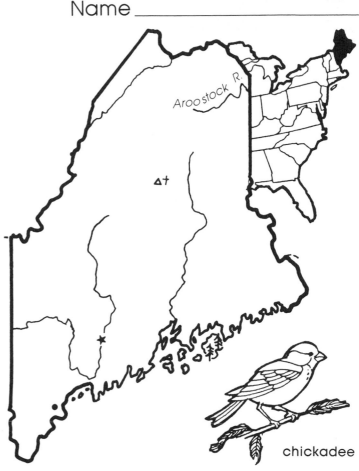

white pine cone and tassel

chickadee

Label the capital, bordering provinces, state, ocean, highest point, national park and Portland (the largest city).

MORE TO LEARN

Write the number of the phase in Column **B** that describes who or what each is in Column **A**. Think logically!

A	B
__ Margaret Chase Smith	1. State capital
__ Mt. Katahdin	2. Only bordering state
__ Henry Wadsworth Longfellow	3. Backbone of the economy
__ Quebec and New Brunswick	4. Potato-producing county
__ New Hamphire	5. Bordering Canadian provinces
__ Augusta	6. Highest point on Atlantic Coast
__ Acadia National Park	7. Famous poet
__ Wood-processing industry	8. Both a U. S. Senator and Congresswoman
__ Aroostook	9. Carter's Secretary of State
__ Edmund Muskie	10. New England's only national park

CAN YOU FIND... Name a famous explorer from Maine.

19

Maryland the Old Line State

Name _____

Maryland, the 7th state, is separated into two parts by the Chesapeake Bay — the nation's largest bay. The bay, which empties into the Atlantic Ocean, has dominated Maryland life for 300 years. It has fertile coastal lands, provides routes for trade and supports a large fishing industry.

The state is bordered on the north by Pennsylvania, on the east by Delaware, on the south by Virginia and on the west by West Virginia. Most of the population is centered around Washington, D. C. and Baltimore.

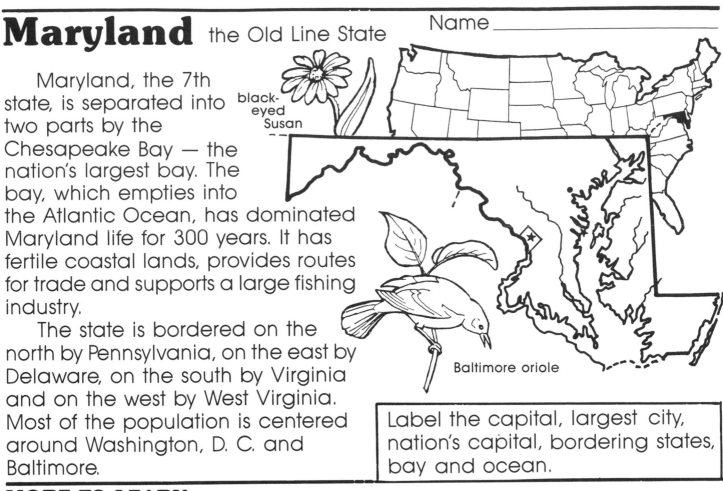

black-eyed Susan

Baltimore oriole

Label the capital, largest city, nation's capital, bordering states, bay and ocean.

MORE TO LEARN

Using the **WORD BANK**, fill in the blanks to learn more about Maryland people and events. The circled letters, when unscrambled, spell the last name of the leader of the Underground Railroad during the Civil War,

Harriet _____.

WORD BANK		
Francis Scott Key	District of Columbia	Baltimore
Spiro Agnew	Annapolis	Mason Dixon Line

_ Ⓞ _ _ _ _ _ _ _, state capital and home of the U. S. Naval Academy

_ _ _ _ _ _ _ _ _ _ _   _ _   _ _ _ Ⓞ _ _ _ _, former Maryland territory

_ _ _ _ _ _ _ _   _ _ _ _ Ⓞ _ _ _ _, writer of "The Star-Spangled Banner"

Ⓞ _ _ _ _ _ _ _ _, largest city in the state

Ⓞ _ _ _ _   _ _ _ _ _ _   _ _ _ _, state's northern boundary

_ _ _ _ _ _   Ⓞ _ _ _ _, U. S. Vice President who resigned

CAN YOU FIND . . . What is the name of the fort that survived the British bombardment and inspired the writing of our national anthem? _____

Massachusetts
the Bay State

Name _____

Massachusetts, the 6th state to join the Union, is also 6th smallest in size. But it stands among leaders in several areas. Boston, the capital, is a major seaport with a growing electronics industry.

The state is an important manufacturing center, leading the nation in production of shoes and leather goods.

Tourists flock to the state for two reasons: the natural attractions such as the Berkshire Mountains, quaint fishing villages, the resort islands of Martha's Vineyard and Nantucket, and the historic attractions centered around Boston with its Bunker Hill Monument, "Old Ironsides" and the "Freedom Trail".

mayflower

On the back of this paper, write the names of the bordering states in **ABC** order.

MORE TO LEARN

Using the **WORD BANK**, fill in the blanks to find out who the first and only Roman Catholic U. S. President was.

WORD BANK					
Boston	Paul Revere	Lexington	Mayflower	Kennedy	Elias Howe

Current U. S. Senator ⦶ _ _ _ _ _ _ _
Inventor of sewing machine _ _ _ _ _ _ _ _ _ _⦶
Where the Revolutionary War began _ _ _ _ _⦶ _ _ _ _
City known as "Cradle of Liberty" _ _ _ _ _ _⦶
Famous "rider" in poem _ _ _ _ _ _⦶ _ _ _ _
Well-known summer resort area _ _ _ _ _ _ _ _⦶
Ship Pilgrims sailed on _ _⦶_ _ _ _ _ _

chickadee

The circled letters spell the last name of John F. _____.

CAN YOU FIND . . . What were three "firsts" in the American Colonies established in Massachusetts in the field of education?

Michigan the Wolverine State

Name _____

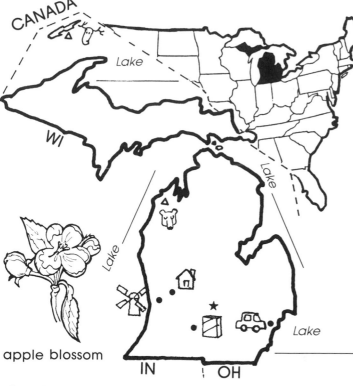

CANADA

Lake

WI

Lake

Lake

apple blossom

IN | OH

Michigan, the 26th state, is often called the Great Lake State because it touches four of them: Lake Superior to the north, Lake Michigan to the west, Lake Huron to the northeast and Lake Erie to the southeast.

Some points of interest are:
- Detroit, the largest city and center of the auto industry
- Battle Creek, the world's largest producer of breakfast cereal
- Holland, site of the only authentic Dutch windmill in the nation
- Grand Rapids, former home of Gerald Ford and the site of his museum
- Lansing, the state capital
- Sleeping Bear Dunes National Lakeshore

The state's only national park, Isle Royale, has one of the largest herds of great-antlered moose left in the U.S.

On the map, label the points of interest and write them below.

★ _____

• _____   • _____

• _____   • _____

🐾 _____   🐾 _____

🐾 _____   🐾 _____

▼ _____   △ _____

robin

MORE TO LEARN

C	A	D	I	L	L	A	C	P
B	X	O	N	P	R	T	A	O
F	O	R	D	A	V	V	N	N
R	L	C	I	T	I	O	A	T
S	D	A	A	B	N	H	D	I
T	S	R	N	O	C	I	A	A
O	N	T	A	R	I	O	D	C
C	H	R	Y	S	L	E	R	V

Locate the following in the **WORD SEARCH**:
- The names of five makes of cars named after famous men in Michigan
- The only Great Lake not bordering Michigan
- The country bordering Michigan
- The two states on Michigan's southern border

CAN YOU FIND . . .

Who founded the city of Detroit? _____

Minnesota the Gopher State

Name _____

Minnesota, the 32nd state, is a land of great natural beauty. Its vast wilderness areas attract hunters, campers, hikers and canoeists.

The state has a reputation for being an area to live in which has a high quality of life with its emphasis on education, small economic units, comprehensive civil service and appreciation of the outdoor life.

Natural resources include the iron ore in the Mesabi Range, thick evergreen forests and fertile soil, which helps with the large production of livestock and livestock products.

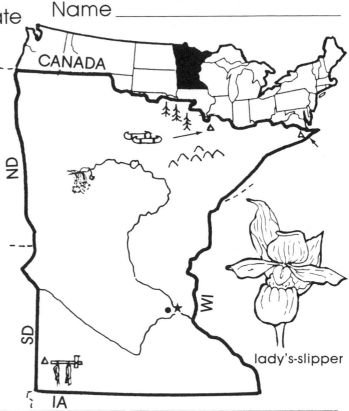

lady's-slipper

Label St. Paul, the capital, Minneapolis, the largest city, national park and monuments, falls and wilderness.

MORE TO LEARN

Unscramble the letters to learn some interesting facts about Minnesota's history, products and natural resources.

Leads the nation in production of these
l o f u r
2 3 1 4 5
and
n e n c d a
3 5 4 1 6 2

g e v e b a t s e l
3 4 1 2 7 6 5 10 9 8
_____ and _____ _____

Two national monuments
a G r d n
3 1 2 5 4
r o P g e t a
3 2 1 6 7 4 5
and

i s t P e p n e o
2 5 6 1 4 3 8 9 7
_____ _____ and _____

Source of the Mississippi River
k e L a
3 4 1 2
t a I s a k
2 3 1 4 6 5
_____ _____

Nation's only wilderness preserved for canoeists
o B d u r a y n
2 1 5 3 7 6 8 4

a W t s r e
2 1 3 6 5 4
_____ _____ Canoe Area

Famous falls in Longfellow's "Song of Hiawatha"
n h h n e M i a a h
3 6 4 5 1 2 9 7 8

_____

National park
o g e V y u r s a
2 5 6 1 3 7 8 9 4
_____

CAN YOU FIND... Who was the legendary lumberjack in Minnesota? _____

loon

Mississippi

the Magnolia State

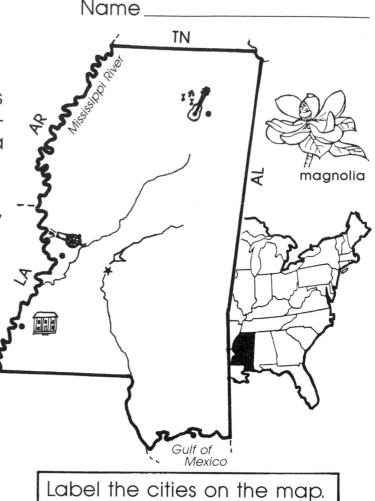

Mississippi, the 20th state, brings to mind pictures of the Old South — cotton plantations, stately magnolia trees and young men in gray.

Today only 25% of the goods produced are agriculturally based, with soybeans the most valuable crop. Manufacturing accounts for 66% with shipbuilding being the leading single industry.

Its cities include: Jackson, the capital and largest city; Vicksburg, site of crucial Civil War Battle; Natchez, site of classic antebellum mansions; Biloxi, Gulfcoast resort area; and Tupelo, Elvis Presley's birthplace.

magnolia

Label the cities on the map.

MORE TO LEARN

Mark each statement **TRUE** or **FALSE**. On another sheet of paper, rewrite each FALSE statement so it is TRUE.

_____ Mississippi is bordered by five states.

_____ Natchez is the site of many antebellum mansions.

_____ The state has no industry.

_____ Louisiana borders on the south.

_____ Biloxi is a well-known industrial center.

_____ Elvis Presley was born in Memphis, Tennessee.

_____ Jackson is the capital and largest city.

_____ Cotton is the most valuable crop.

_____ Shipbuilding is the leading single industry.

_____ Mississippians fought in the Union army.

mockingbird

CAN YOU FIND . . . What was the name of the last home of Jefferson Davis, President of the Confederacy? _____

Missouri
the Show Me State

Missouri, the 24th state, is a land of great natural beauty. Bordering on the east is the mighty Mississippi River. Flowing eastward from Kansas City to St. Louis is the Missouri River, the source of many of Mark Twain's stories.

In the Ozark Mountain area is the beautiful Lake of the Ozarks. The lake has more than 1,300 miles of shoreline, nearly as much as huge Lake Michigan.

Because of competition between St. Louis and Kansas City, the capital is located in the center of the state at Jefferson City.

bluebird

Label the points of interest on the map and write them below.

★ _____

● _____   ● _____

{ _____   { _____

⋀ _____   ◠ _____

hawthorne

MORE TO LEARN

Missouri and Tennessee are the only states bordered by eight other states. Using the directional clues below, label the bordering states on the map. Use postal abbreviations.

Iowa—(n)orth   Arkansas—sou(t)h
Kansas—west   Oklahoma—southwest
Illinois—e(a)st   Nebraska—no(r)thwest
Kentucky—so(u)theast
Tennessee—the (m)ost southeast

Unscramble the circled letters to find out the last name of the only U.S. President from Missouri.

_____

CAN YOU FIND . . . Who were the French explorers that discovered the area which is Missouri? _____ _____

Montana
the Treasure State

western meadow lark

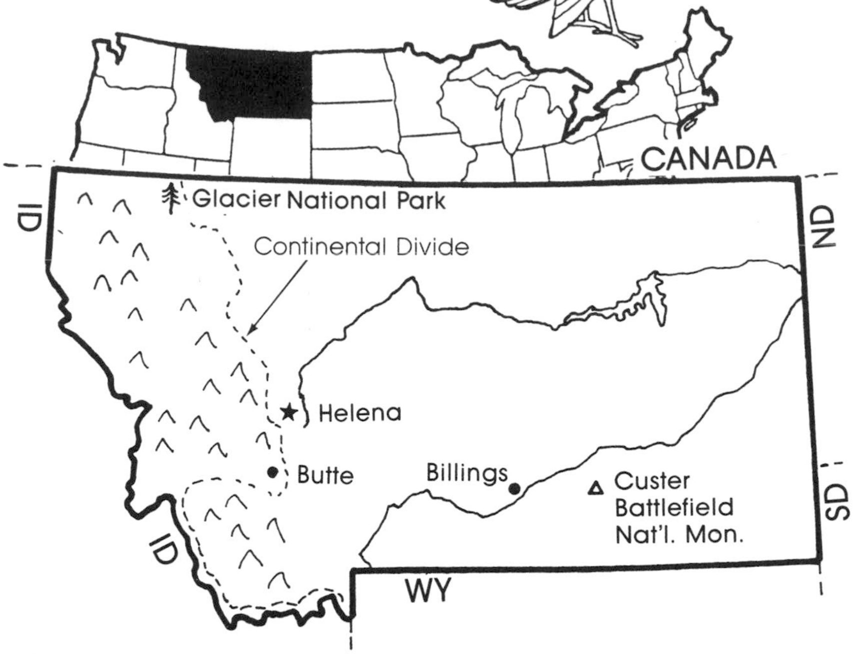

CANADA

^ ^ Glacier National Park

Continental Divide

ID

★ Helena

• Butte   Billings •   △ Custer
                              Battlefield
                              Nat'l. Mon.

ND

SD

WY

Montana, the 41st state, got its name from a Latin word meaning "mountainous". Located in the mountains are rich mineral deposits. The discovery of gold brought the first settlers — long after Lewis and Clark explored the region in 1805. Later, the copper deposits became more important. Today, petroleum is the state's leading mineral product. However, agriculture is one of the biggest money-makers, with large ranches and farms.

Its Glacier National Park, which borders Canada, has mountains never climbed and, perhaps, treasures to be discovered.

Fill in the blanks below using the symbols as clues

★ _____   • _____

• _____   ♣ _____

- - - _____

△ _____

bitterroot

MORE TO LEARN

Country bordering Montana
Large economic moneymaker
First American explorers in the region
Montana means . . .
State capital
Minerals which influenced state's early history

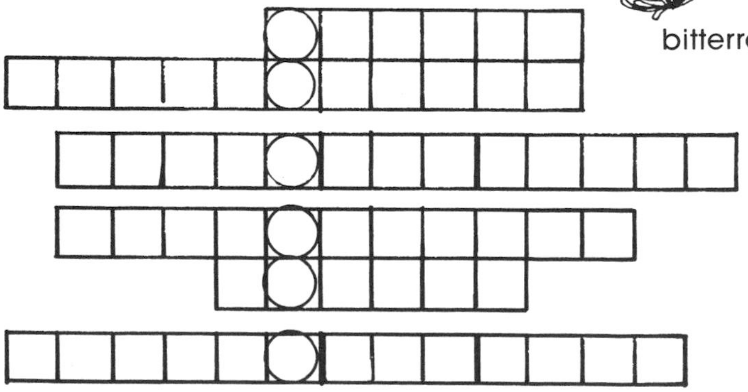

The circled letters spell the last name of the colonel who led the battle at Little Bighorn. _____

CAN YOU FIND . . . Who was the Montana man who was a legend in Congress for thirty-four years? _____

Nebraska

the Cornhusker State

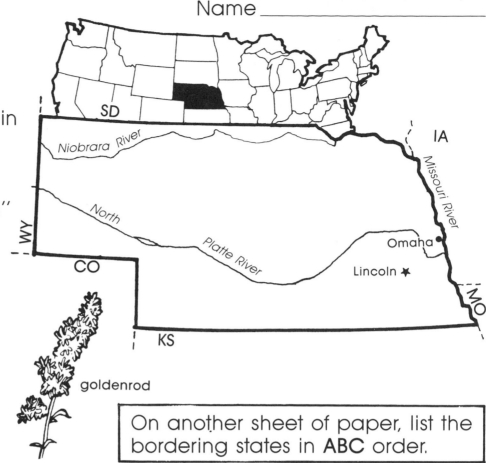

Name _____

Nebraska, which became the 37th state in 1867, is known for the hardy, pioneer spirit of the people.

The first "homestead" in the nation was claimed in 1862, ten years before J. Sterling Morton founded Arbor Day. When the area became the Nebraska Territory in 1854, it was nearly treeless. Now it has the only national forest in the nation planted by foresters.

goldenrod

> On another sheet of paper, list the bordering states in **ABC** order.

In 1937 Nebraska adopted the only unicameral (one house) legislature in the nation, and in 1913 Leslie King, Jr. was born. Today he is known as Gerald Ford!

MORE TO LEARN

Number the historic events listed below in the order they occurred.

___ Nebraska (b)ecame a territory.
___ A (u)nicamera(l) (l)egislature was adopted.
___ J. Sterl(i)ng Morton (f)ounded Arbor Day.
___ Ne(b)raska became the 37th state.
___ Gerald Ford was born.
___ The (f)irst "h(o)mestead" w(a)s c(l)aimed.

western meadow lark

Write the circled letters. _____ When unscrambled, they spell the name of the man who organized a famous Wild West Show.

_ _ _ _ _ _ _ _ _ _

CAN YOU FIND . . . What archeological find was unearthed near North Platte in 1922? _____

Nevada

the Silver State

Nevada, the 36th state, is a land of rugged beauty with its snow-capped mountains, sandy deserts and jagged plateaus.

Less rain falls in Nevada than in any other state. What brought people to the state was the rich mineral deposits — miners who wanted to strike it rich. Today 20 million tourists came to strike it rich — in the gambling casinos. Eighty percent of the state's residents live in or near Las Vegas and Reno and are dependent on the gambling industry for their livelihood.

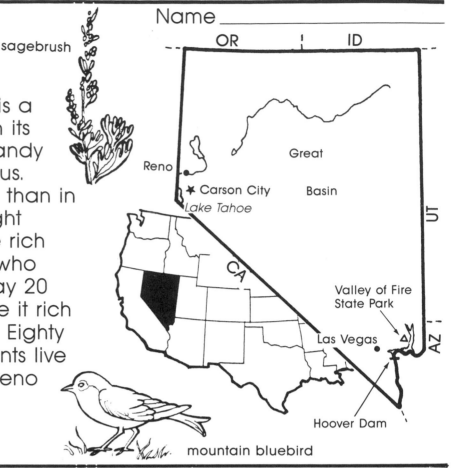

sagebrush

mountain bluebird

MORE TO LEARN

Write the number of the phase in Column **B** that describes who or what each is in Column **A**. Think logically!

A	B
__ Comstock Lode	1. Formed Lake Meade, a huge man made lake
__ Tourism	2. Large desert area
__ Valley of Fire	3. State park with unusual rock formations
__ Hoover Dam	4. In 1859 a rich deposit of gold and silver
__ Lake Tahoe	5. State's largest city and gambling center
__ Las Vegas	6. President when Nevada admitted to the Union
__ Abraham Lincoln	7. Biggest industry in the state
__ Carson City	8. State capital
__ Great Basin	9. Beautiful lake resort area

CAN YOU FIND . . . What was the number of residents required for a territory to become a state in 1864? _____ How many residents were in Nevada in 1880? _____

New Hampshire

the Granite State

New Hampshire, the 9th state, is well known for its natural beauty. It is entirely bordered by Vermont on the west, by Massachusetts on the south, Canada on the north and Maine (and the Atlantic Ocean) on the east. Its Mt. Washington is New England's highest peak.

Though not important in New Hampshire's economy, its granite was used to build the Library of Congress and to make the cornerstone for the United Nations building.

A variety of people from the state played important roles in U.S. history: Franklin Pierce, the 14th President; Mary Baker Eddy, founder of the Christian Science movement; and Alan Shepard, first American astronaut in space.

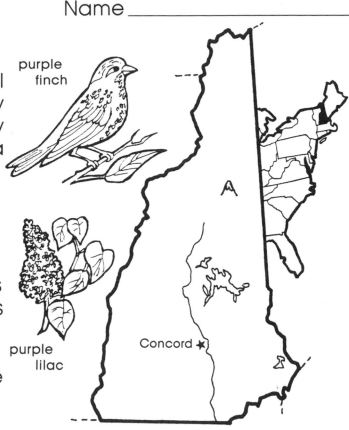

purple finch

purple lilac

Concord ★

Label the bordering states, country, ocean and the highest peak.

MORE TO LEARN

Mark each statement **(T) TRUE** or **(F)FALSE**. On another sheet of paper, rewrite each FALSE statement so it is TRUE.

_____ Granite is important to New Hampshire's economy.

_____ Mt. Washington is New England's highest peak.

_____ Maine is bordered by only one state — New Hampshire.

_____ Franklin Pierce was the 9th U.S. President.

_____ Vermont borders the state on the east.

_____ Alan Shepard was the first American to walk in space.

_____ New Hampshire granite was used to build the Library of Congress.

_____ Mary Baker Eddy founded the Christian Scientist movement.

_____ New Hampshire was the 8th state to ratify the Constitution.

CAN YOU FIND . . . Who was the famous orator and senator in the 1800's?

_____

29

New Jersey
the Garden State

Name _____

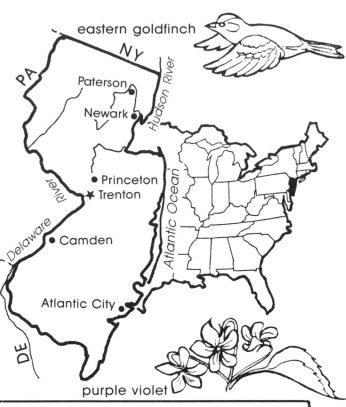

eastern goldfinch
PA
NY
Paterson
Newark
Hudson River
Princeton
Trenton
Delaware River
Camden
Atlantic Ocean
Atlantic City
DE
purple violet

New Jersey, the 3rd state, has been greatly influenced by its location. It lies between the Hudson and Delaware Rivers and along the Atlantic Coast. Products made there can be shipped throughout the U.S. and other countries.

Located between two giant cities, New York City and Philadelphia, the state has a huge nearby market for its farm products. Thousands of New Jerseyans work in these two cities and commute.

Located on the coast are many resort areas. Atlantic City is the most famous.

> On another sheet of paper, list the major cities in **ABC** order.

MORE TO LEARN

Unscramble the letters to learn about some interesting people and events in New Jersey history.

a s T h m o   d E s i o n
5 6 1 2 4 3   2 1 4 3 5 6
_____ invented the electric light in his lab at Menlo Park.

m a u l S e   o M r e s
3 2 4 6 1 5   2 1 3 6 5
_____ developed the electric telegraph near Morristown.

v G o r e r   l e v C e n d a l
4 1 3 2 5 6   2 3 4 1 5 8 9 7 6
_____ , 22nd U.S. President

d o W o r o w   I s n W i o
4 3 1 2 5 6 7   3 4 6 1 2 5
_____ , 28th U.S. President

b o H o k n e
3 4 1 2 5 7 6
_____ , site of the first pro baseball game in 1846

r t e T o n n   r c i e P n n o t
2 5 3 1 6 7 4   2 5 3 6 1 4 9 8 7
_____ and _____ , cities that served as the nation's capital

CAN YOU FIND . . . Where and when was the first dinosaur skeleton found in North America? _____

New Mexico

the Land of Enchantment

New Mexico, the 47th state, is rich in history, beauty and natural resources. Founded in 1610 by the Spanish, Santa Fe is the oldest seat of government in the nation, and El Camino Real is the oldest road.

The Indian influence can be seen in such names as the Navajo and Elephant Butte Dams and at such events as the Inter-Tribal Indian Ceremonial held each year.

A land of contrasts, New Mexico boasts the nation's greatest source of uranium ore (the first atomic bomb was exploded near Alamogordo in 1942), the beautiful White Sands desert area and Carlsbad Caverns, with the world's largest underground room.

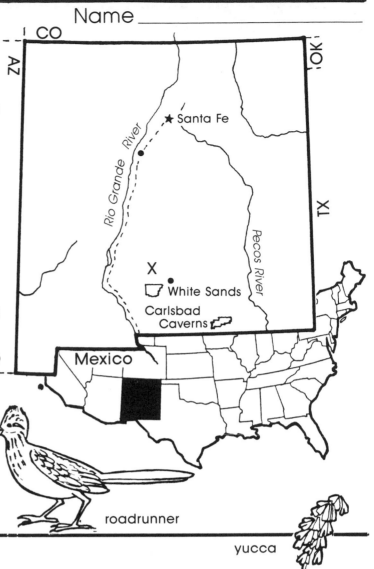

roadrunner

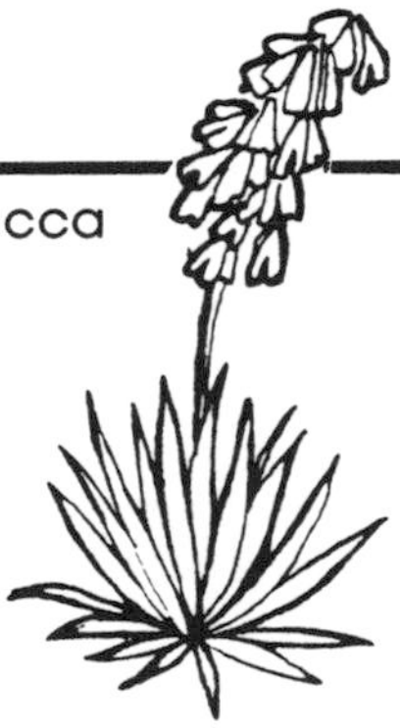

yucca

MORE TO LEARN

N _____ Dam created new farmland.

E _____ forms the state's largest lake.

W _____ , a large deposit of gypsum sand and a national monument

M ineral contributing to atomic research development, _____

E _____ , nation's oldest road

X marks the spot near _____ of the first atomic bomb explosion.

I _____ influence can be seen throughout the state.

C _____ , one of the world's great natural wonders

O ldest seat of government in the nation, _____

CAN YOU FIND . . . Who found uranium in 1950 in the northwest region of New Mexico? _____

New York

the Empire State

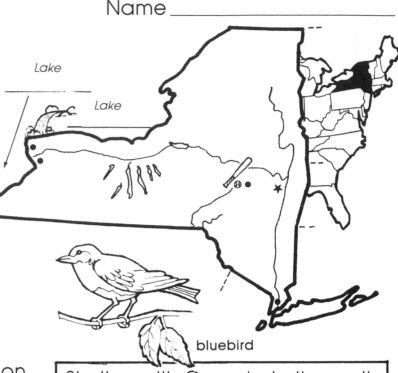

New York, the 11th state, has earned the nickname which came from a remark by George Washington predicting it might become the "seat of the new empire". And in a way, it has.

New York leads the nation in foreign trade and wholesale trade and is second in retail trade, manufacturing and population.

There is New York City, the nation's largest city, and center for banking, trade, communication, finance and transportation. As headquarters for the United Nations, it could be called the "capital of the world"!

Name_____

Lake _____

Lake

bluebird

Starting with Canada to the north, moving clockwise, label Vermont, Massachusetts, Connecticut, New Jersey, Pennsylvania, Lakes Erie and Ontario.
Label New York City, Albany (state capital), Cooperstown, Buffalo, Niagara Falls and the Erie Canal.

MORE TO LEARN

rose

Use the **WORD BANK** to fill in the blanks.

WORD BANK			
Roosevelt	West Point	Statue of Liberty	Martin
Elmira	Millard Fillmore	Cooperstown	Van Buren

Location of the Baseball Hall of Fame _ _ _ _ _ _ _ _ _ _ _ _

Last name of only cousins to be U.S. Presidents _ _ _ _ _ _ _ _ _

U.S. Military Academy _ _ _ _ _ _ _ _ _

Location of Mark Twain's grave _ _ _ _ _ _

Monument in New York Harbor _ _ _ _ _ _ _ _ _ _ _ _ _ _ _

Two other U.S. Presidents from New York _ _ _ _ _ _ _ _ _ _

_ _ _ _ _ and _ _ _ _ _ _ _ _ _ _ _ _ _ _ _ _

CAN YOU FIND . . .

How much was paid to the Indians for Manhattan Island?

_____

North Carolina
the Tar Heel State

North Carolina, the 12th state to ratify the Constitution, leads the nation in tobacco farming and production of tobacco products. It also makes more wooden furniture and more cloth than any other state.

There are some beautiful, scenic areas in the Blue Ridge Mountain region which includes the Great Smoky Mountains and some lovely waterfalls. Pinehurst is a famous winter resort area in the western coastal plain.

MORE TO LEARN

Number the cities, points of interest, bordering states and bordering bodies of water shown on the map.

___ Wright Brothers Monument at Kitty Hawk

___ Raleigh, the state capital

___ Great Smoky Mountains

___ Pamlico Sound

___ Fayetteville, on the Cape Fear River

___ Greensboro, world's largest mill for weaving denim

___ Kannapolis, world's producer of household textiles, such as towels

___ Mt. Mitchell, eastern American's highest peak

___ Charlotte, the largest city

___ Atlantic Ocean

___ Virginia, to the north

___ Tennessee, to the west

___ South Carolina, to the south

___ Georgia, to the southwest

___ Cape Hatteras, site of many shipwrecks

flowering dogwood

cardinal

CAN YOU FIND . . .   What was Blackbeard the Pirate's real name?

_____

North Dakota
the Flickertail State

North Dakota, the 39th state, has a higher percentage of people working in some form of agriculture than any other state. Only 18 cities have more than 2,500 people. The farmers have fought hard to keep cooperatives taking over from families.

Wheat is grown in every county. But North Dakota is also the nation's leader in production of barley, sunflower seeds and flaxseed.

Long ago the Sioux named the land "Dakota" meaning friends. That has modern significance in that the state built an International Peace Garden to commemorate the friendship between the U.S. and Canada.

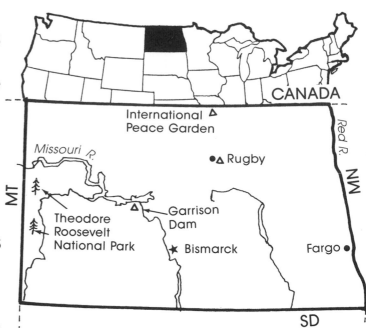

Write the names of the following cities or points of interest.

★ _____ • _____

🌲 _____ △ _____

△ _____ •△ _____

MORE TO LEARN

Write the number of the phrase in Column **B** that describes who or what each is in Column **A**. Think logically!

western meadowlark

A

___ The only bank
___ North Central Rugby
___ −60° to 121°
___ Soil
___ International Peace Garden
___ Treaty with Sioux
___ Two SAC bases
___ Sacagawea

wild prairie rose

B

1. The state's most precious resource
2. Symbol of enduring friendship with Canada
3. Brought peace to the area
4. Located in North Dakota
5. Owned by a state
6. Famous Indian guide to Lewis and Clark
7. Geographic center of North America
8. U.S. record for one year temperature range

CAN YOU FIND . . . Whose cabin is located on the grounds of the state capital?

_____

Ohio

the Buckeye State

cardinal

scarlet carnation

Ohio, the 17th state, comes from the Indian word meaning "something great". The state has earned the name! Though 35th in size, it ranks 6th in population and 3rd in value of goods produced.

Some of the factors that have contributed to its growth and development are: its central location, its abundant water and fertile soils, and its valuable mineral resources.

Last, but not least, it was the home of seven U.S. Presidents!

Name _____

Label the capital, cities and bordering states: Michigan — north, Indiana — west, Pennsylvania — northeast, West Virginia — southeast, Kentucky — south.

MORE TO LEARN

Circle the locations and some of Ohio's famous people in the **WORD SEARCH**.

```
T V R B C X P N A
H A K R O N D E R
O B E R L I N I M
M E J G U M A L S
A D O G M C L C T
S I H E B G E C R
A S N E U F V A O
K O N N S F E N N
R N I A N F L T G
O B E R L E C O Z
F X L M S Y V N M
```

- Thomas Edison, inventor of the lightbulb
- John Glenn, U.S. Senator and former astronaut
- Canton, location of Pro Football Hall of Fame
- Oberlin, first coed college in the U.S.
- McGuffey, author of famous old readers
- Neil Armstrong, first astronaut on the moon
- Columbus, the capital
- Akron, world's greatest rubber producer
- Cleveland, largest city (near northern border)

CAN YOU FIND . . . Who were the seven U.S. Presidents born in Ohio? _____

_____

Oklahoma the Sooner State

Name _____

Oklahoma is an Indian word meaning "red people". In the 1800's the land was set aside as a huge Indian reservation for "as long as grass shall grow and rivers run". Today only about 4 percent of the population is Indian.

The Five Civilized Tribes wanted to become the state of Sequoyah in 1905. But Congress refused. So in 1907, the Oklahoma and Indian Territories united to become the 46th state.

The "Dust Bowl" of the 1930's is fertile land again for farming and raising cattle, and the discovery of oil brought a new source of wealth.

CO

Land of the Five Civilized Tribes in the 1820's → Cherokee, Creek, Seminole, Chickasaw, Choctaw

Beginning with Colorado, and moving clockwise, label the bordering states: Kansas, Missouri, Arkansas, Texas and New Mexico.

scissor-tailed flycatcher

MORE TO LEARN

Use the WORD BANK to fill in the blanks below.

Site of Ft. Sill, the Army's main artillery school _ _ _ _ _ _

Famous humorist, the late _ _ _ _   _ _ _ _ _ _

Choctaw chief who named the territory _ _ _ _ _   _ _ _ _ _

What the Indian journey is often called _ _ _ _ _   _ _   _ _ _ _ _

A Cherokee Confederate brigadier general _ _ _ _ _   _ _ _ _ _

The first capital of Oklahoma _ _ _ _ _ _ _

The present capital of Oklahoma _ _ _ _ _ _ _ _   _ _ _ _

Cherokee leader who invented a writing system _ _ _ _ _ _ _ _

mistletoe

WORD BANK			
Sequoyah	Allen Wright	Trail of Tears	Guthrie
Oklahoma City	Will Rogers	Stand Watie	Lawton

CAN YOU FIND . . .   Why is Oklahoma called the "Sooner State"? _____

36

Oregon the Beaver State

Oregon, the 33rd state, is known for its huge evergreen forests and rugged natural beauty. The mighty Columbia River on the north separates Oregon from Washington. The winding Snake River forms much of its boundary with Idaho on the east. Its beautiful shoreline on the Pacific Ocean is breathtaking. California and Nevada border on the south.

Portland, nicknamed the "City of Roses", is the largest city. Eugene ranks second. Salem is the capital. Crater Lake, Hells Canyon and the Columbia River Gorge help attract millions of tourists each year.

Oregon grape

Label the bordering states, bordering bodies of water, cities and points of interest on the map. (13 items)

western meadowlark

MORE TO LEARN

Use the rhyming clues to fill in the blanks.

Oregon's <u>m</u>ost valuable <u>c</u>rop, _____ (rhymes with cheat)
Resource that is one of the state's most val<u>u</u>able, _____ (rhymes with daughter)
Explorer wh<u>o</u> named Col<u>u</u>mbia River, Robert _____ (rhymes with pray)
Great, year-round skii<u>n</u>g at Mount _____ (rhymes with good)
Oregon's most valuable <u>i</u>ndustry, _____ processing (rhymes with hood)
Natio<u>n</u>'s deepest lake, _____ Lake (rhymes with greater)
The underlined letters, written in order, spell the name of the "Father of Oregon", John _____.

CAN YOU FIND . . . What U.S. President based his campaign slogan (Fifty-Four Forty or Fight!) on the Oregon boundary dispute with the British? _____

Pennsylvania

ruffed grouse

Name _____

the Keystone State

Pennsylvania, the 2nd state to ratify the Constitution, was nicknamed "the Keystone State" because it was the center, or keystone, of the arch of the original 13 colonies. The Declaration of Independence was signed in Philadelphia.

The state today is experiencing difficulties in overcoming its dependence on three troubled industries — coal mining, steel production and railroading. Plus it has had to overcome the disaster in 1979 at the Three Mile Island Nuclear Power Plant. But Pennsylvania is in an era of renewal!

Ohio River

• Hershey

Label the bordering states and Great Lake starting with New York on the north, moving clockwise: New Jersey, Delaware, Maryland, West Virginia, Ohio, Lake Erie. Label Harrisburg, the capital, Philadelphia and Pittsburgh on the Ohio River.

MORE TO LEARN

Write the number of the name in Column **B** with the phase it matches in Column **A**. Think logically!

A	B
___ World's largest chocolate factory	1. William Penn
___ Bordering Great Lake	2. Pittsburgh
___ Declaration of Independence author	3. Hershey
___ The founder of Pennsylvania	4. Thomas Jefferson
___ Scientist, writer, philosopher, inventor	5. England
___ Largest city, cultural and industrial center	6. Philadelphia
___ America's steel capital	7. Benjamin Franklin
___ Country where the Liberty Bell was made	8. Erie

mountain laurel

CAN YOU FIND . . . Where is the location for the "official" groundhog for Groundhog Day? _____

Rhode Island

Little Rhody

Rhode Island, the 13th state, may be the tiniest (48 miles long and 37 miles wide), but this state has had many firsts. Founded by Roger Williams in 1636 in his quest for religious freedom, the state had the first Baptist Congregation and the first Jewish Synagogue in the nation.

Rhode Island was also the location of other American firsts: first dry goods store, first cotton mill, first power loom, first torpedo boat . . . and, Rhode Island ranks first in the world in the production of costume jewelry and sterling silver products.

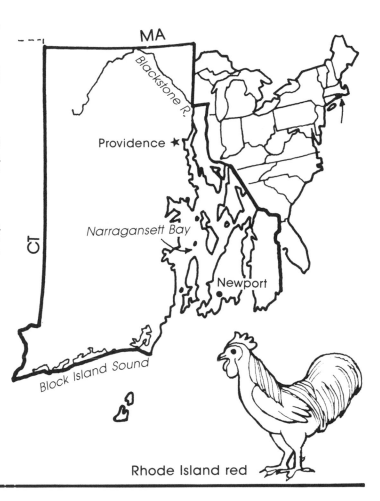

violet

MA

Blackstone R.

Providence ★

CT

Narragansett Bay

Newport

Block Island Sound

Rhode Island red

MORE TO LEARN

ACROSS

1. _____ Bay
5. State capital
7. Made 1st _____ boat
8. State bordering on north

DOWN

1. City in southeast
2. First dry _____ store
3. _____ in size
4. Founder, _____ Williams
6. _____ jewelry center

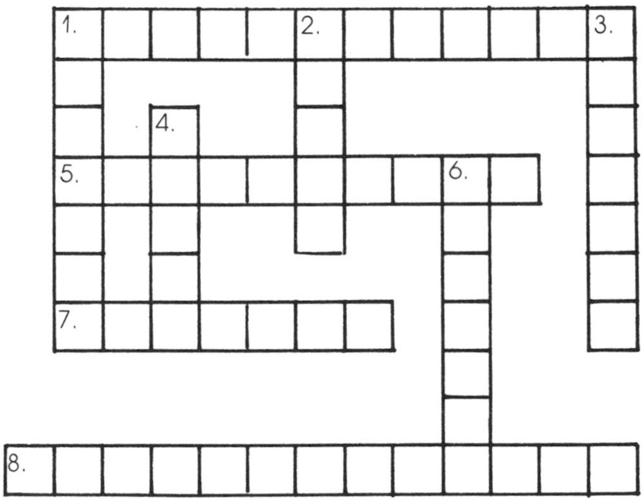

CAN YOU FIND . . . Who is the man from Rhode Island who is the oldest person ever to serve in Congress? _____

South Carolina

the Palmetto State

Name _____

South Carolina was the 8th state to ratify the Constitution and the first state to secede from the Union. The battle that began the Civil War was at Fort Sumter. Sherman burned the capital city, Columbia.

The state is bounded by North Carolina on the north, Georgia on the west and the Atlantic Ocean on the east. Charleston, in the coastal lowlands, is its most historic city with an Old South flavor. Greenville, in the northwest is the third largest city, after Columbia and Charleston.

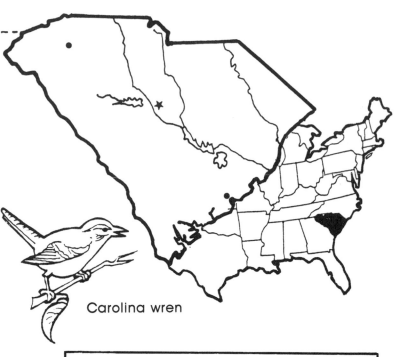

Carolina wren

Label the capital, large cities, bordering states and ocean.

MORE TO LEARN

Fill in the blanks from the **WORD BANK**

WORD BANK

tobacco	Calhoun	Santee	Citadel
	Myrtle Beach	Parris Island	

Famous seaside resort ○_ _ _ _ _ _ _ _ _ _

Vice President under Jackson, John _○_ _ _ _ _ _

U.S. Marine base _ _○_ _ _ _ _ _ _ _ _

Military school at Charleston _○_ _ _ _ _ _

Leading cash crop _○_ _ _ _ _ _

Large river in the state _ _○_ _ _

Carolina jessamine

The circled letters spell the last name of a Revolutionary War hero known as the Swamp Fox, Francis _____.

CAN YOU FIND . . . Who was the first Republican governor in 100 years elected in 1974? _____

40

South Dakota

Name _____

the Sunshine State (Also known as the Coyote State.)

South Dakota, the 40th state, is a "Land of Infinite Variety" with its rich farmlands east of the wide Missouri River and its deep canyons and Badlands west of the river. Located in the Black Hills is Mount Rushmore, with the heads of four Presidents carved in its granite, and Homestake, the largest gold mine in the Western Hemisphere.

Sioux Falls, the largest city, gives evidence to the great place Indians had in the history of the state's development. A large monument is being built to Chief Crazy Horse.

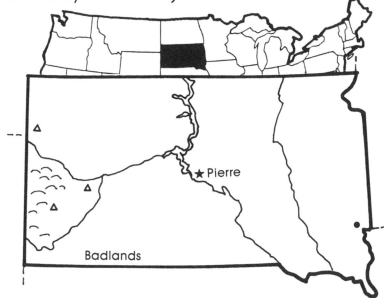

★ Pierre

Badlands

Chinese ring-necked pheasant

Label the following on the map: the largest city, the Black Hills, Mount Rushmore, Homestake, the river in the center, the geographic center of the U.S. and the bordering states — North Dakota–north, Iowa–southeast, Nebraska–south, Montana–northwest, Wyoming–southwest, Minnesota–east

MORE TO LEARN

H	S	I	O	U	X	F	A	L	L	S
O	M	A	M	V	R	B	L	E	J	R
M	O	N	I	P	I	E	R	R	E	O
E	N	A	S	N	L	T	H	V	F	O
S	B	T	S	A	I	C	H	R	F	S
T	L	H	O	P	N	R	O	B	E	E
A	A	I	U	O	C	A	R	M	R	V
K	C	L	R	T	O	Z	S	E	S	E
E	K	L	I	S	L	Y	E	T	O	L
W	A	S	H	I	N	G	T	O	N	T

Locate the following in the **WORD SEARCH**: (They read → or ↓).
The four Presidents carved on Mount Rushmore
The capital and largest city
Famous Indian Chief
Location of Mount Rushmore
Name of the famous gold mine
Wide river in state's center

pasqueflower

CAN YOU FIND . . . What famous writer of children's books is from South Dakota?

_____

Tennessee
the Volunteer State

mockingbird

Mississippi R.

Tennessee R.

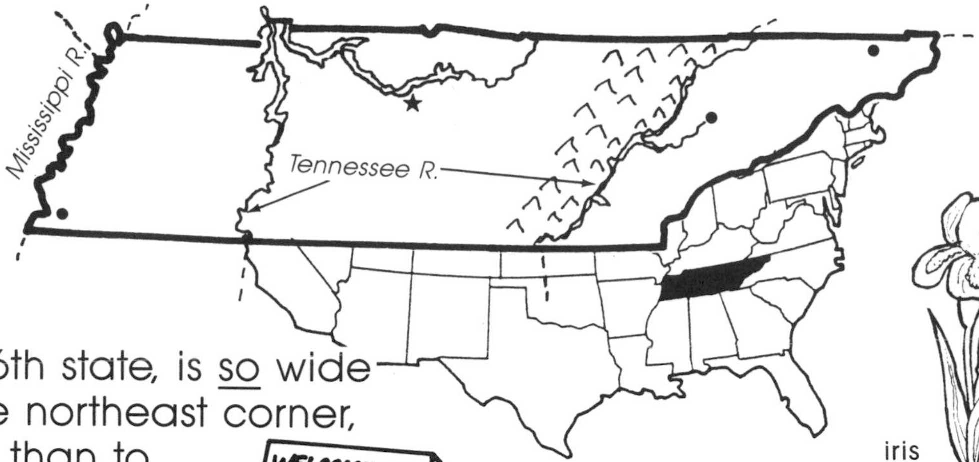

iris

Tennessee, the 16th state, is <u>so</u> wide that Kingsport, in the northeast corner, is closer to Canada than to Memphis. And Memphis is in the southwest corner of the state! Many highway signs say . . .

WELCOME TO THE 3 GREAT STATES OF TENNESSEE

East Tennessee has the beautiful Great Smoky Mountains and the Blue Ridge Mountains along the eastern edge. Knoxville is the largest city. Middle Tennessee has rolling hills, beautiful horses and Nashville — the state capital and home of country music.

West Tennessee has flat, fertile farmland and Memphis, the largest city in the state.

Label the points of interest on the map and write them below.

★ _____
• _____
• _____
• _____
⌃⌃ _____
⌃⌃ _____

MORE TO LEARN

Tennessee and Missouri are the only states bordered by eight other states. Using the directional clues below, label the bordering states on the map. Use postal abbreviations.

Virginia–north at the top east side
Kentucky–north
Missouri–northwest
Arkansas–southwest
Mississippi–south on the west side
Alabama–south central
Georgia–southeast
North Carolina–east

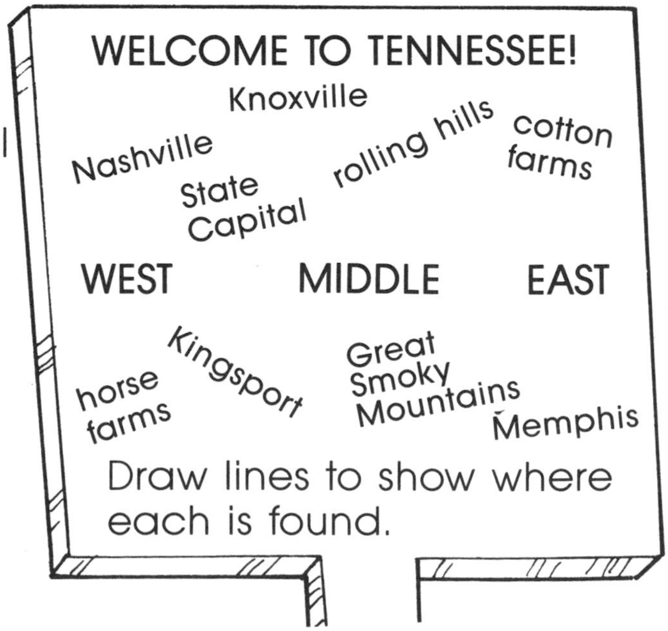

WELCOME TO TENNESSEE!

Knoxville

Nashville   State Capital   rolling hills   cotton farms

WEST   MIDDLE   EAST

horse farms   Kingsport   Great Smoky Mountains   Memphis

Draw lines to show where each is found.

CAN YOU FIND . . .   Who was the U.S. President from Tennessee?

_____

Texas the Lone Star State

Name _____

Texas, the 28th state, is big!! Big in size, big in population, big in products and big in history. Second only to Alaska in size, its King Ranch is bigger than Rhode Island!

The state has the most farms, farmland, cattle, horses and sheep in the nation. It leads the nation in the production of oil, natural gas and electrical power. It is the greatest U. S. source of salt, magnesium and sulphur.

The well-known cities are Houston, with the Manned Space Flight Center, Dallas, in the heart of the oil and cotton region, Ft. Worth, the cattle capital, and Austin, the capital.

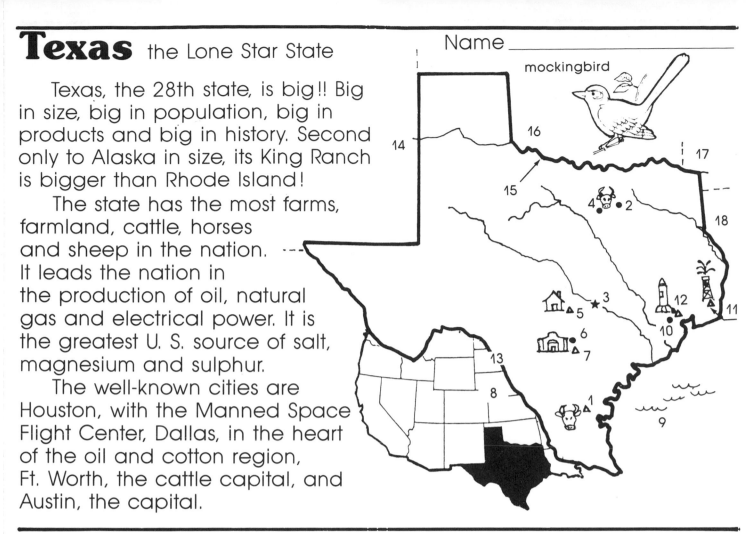

mockingbird

bluebonnet

MORE TO LEARN

Number the cities, points of interest, bordering states, country, rivers and body of water shown on the map.

___ Mexico, on the southwest

___ The famous Alamo in San Antonio

___ King Ranch

___ Manned Space Flight Center

___ Rio Grande River, bordering Mexico

___ Red River, bordering Oklahoma

___ New Mexico, on the west

___ Louisiana, on the southeast

___ Spindletop Monument, site of gusher oil

___ Ft. Worth

___ Dallas

___ Austin

___ San Antonio

___ Houston

___ Oklahoma, on the north

___ L. B. Johnson Home

___ Arkansas, on the northeast

___ Gulf of Mexico

CAN YOU FIND . . . Who was the U. S. President assassinated in Dallas?

_____

43

Utah

the Beehive State

Name _____

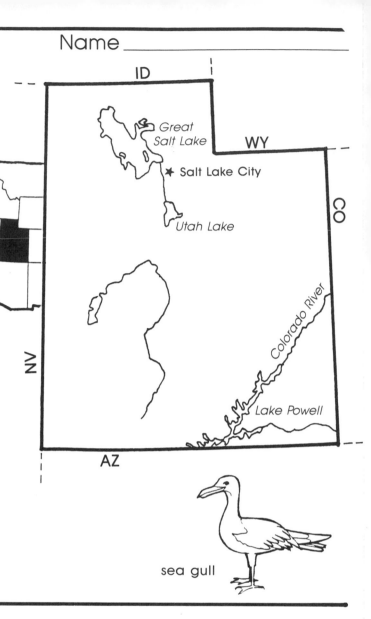

"This is the place!" said Brigham Young in 1847 when he led Mormon settlers into the Salt Lake Valley to escape religious persecution. A year later the U.S. acquired the territory from Mexico. In 1850 it became the Utah Territory, a year after Young established the state of Deseret. Because of the Mormon practice of polygamy, it did not become the 45th state until 1896 (six years after suspending the practice). In fact, it even led to the Utah War in 1857.

The state bird is the sea gull because, in 1848, these birds miraculously appeared and destroyed swarms of grasshoppers which were devouring the crops!

sea gull

MORE TO LEARN

Number the historic events listed below in the order they occurred.

___ Utah (j)oined the Uni(o)n as the 45th state.

___ The U.(S.) Army occupied the territory in the Utah War.

___ S(e)a gulls saved the Mormon's cro(p)s.

___ T(h)e Mormons suspended the practice of polygamy.

___ The U.(S.) acquired the territory from Mexico.

___ Brigha(m) Young led Mormon settlers to the territory.

___ The area became the Utah Terr(i)tory.

___ Young es(t)ablished t(h)e state of Deseret.

sego lily

The circled letters spell _____, founder of the Mormon Church.

CAN YOU FIND . . . What is the name of the ski resort developed by actor Robert Redford? _____

Vermont
the Green Mountain State

Vermont was the first state to join the newly-formed United States in 1791. The area had been awarded by the King of England to New York in 1764. But Ethan Allen and the Green Mountain Boys captured Fort Ticonderoga in 1775. An independent republic was established and a constitution adopted in 1777.

This constitution provided for state education, the right for every male to vote, and it forbid slavery. All firsts in America!

With its beautiful mountain scenery, Vermont is a great tourist area throughout the seasons.

Fill in the blanks.

★ _____

● _____

⸹ _____

⸹ _____

⸹ _____

red clover

hermit thrush

MORE TO LEARN

Mark each statement **TRUE** or **FALSE**. On another sheet of paper, rewrite each FALSE statement so it is TRUE.

_____ Vermont is very flat.

_____ The state attracts tourists year round.

_____ From its early history, Vermont has believed in the rights of the individual

_____ The original state constitution gave all adults the right to vote.

_____ Vermont was the 14th state admitted to the Union.

_____ The King of France awarded Vermont to New York.

_____ Vermont has always forbidden slavery in its constitution.

_____ The Green Mountain boys lost the battle at Fort Ticonderoga.

CAN YOU FIND . . .   What two U.S. Presidents were born in Vermont?

_____

Virginia Old Dominion

Name _____

Virginia, the 10th state, is more steeped in history than perhaps any other state. The list of important events is endless!

Nicknamed Old Dominion, Virginia is also known as "Mother of Presidents" as eight were born there and "Mother of States" as all or part of eight states were formed from it.

The greatest documents of freedom came from Virginians; the Declaration of Independence, the Constitution and the Bill of Rights.

cardinal

Arlington

Richmond ★
Williamsburg
Jamestown

Label the bordering states, bay and ocean, starting on the north and moving clockwise: Maryland, Chesapeake Bay, Atlantic Ocean, North Carolina, Tennessee, Kentucky and West Virginia.

MORE TO LEARN

Using the **WORD BANK**, fill in the blanks to learn about some historic events, places and people in our nation's history.

WORD BANK	Mt. Vernon Monticello Patrick Henry	Pentagon Appomattox Robert E. Lee	Williamsburg Jamestown	Arlington Thomas Jefferson

_ _ _ _ _ _ _ _ _ _ _ , Thomas Jefferson's home

_ _ . _ _ _ _ _ _ _ , George Washington's home

_ _ _ _ _ _ _ _ _ _ _ , site of Confederate surrender

_ _ _ _ _ _ _ _ _ , first permanent English settlement in America

_ _ _ _ _ _ _ _ _ _ _ _ _ _ _ _ _ , author of the Declaration of Independence

flowering dogwood

_ _ _ _ _ _ _ _ _ , our National Cemetery

_ _ _ _ _ _ _ _ _ _ _ _ _ , restored colonial town

_ _ _ _ _ _ _ _ _ _ _ _ _ , said, "Give me liberty or give me death."

_ _ _ _ _ _ _ _ _ , the massive military center

_ _ _ _ _ _ _ . _ _ _ _ , the outstanding Confederate general

CAN YOU FIND . . .	Where and when did the British surrender to end the last major battle of the Revolutionary War? _____

Washington
the Evergreen State

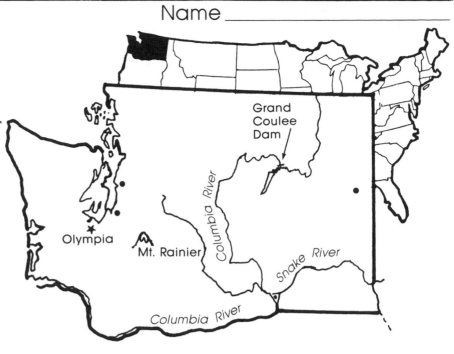

Name _____

Washington, the 42nd state, is a land of contrasts. The Cascade Mountains divide the state in two economic and geographic regions. In the east is one of the nation's most productive farm areas with its marketing center in Spokane. In the western lowlands are the industrial centers of Seattle and Tacoma. In the east are high mountains and thick forests. In the west are flat, treeless desert lands.

Water is one of its most important resources as it has more potential water power than any other state. The Grand Coulee Dam, the largest concrete dam in the U.S., is one of the world's greatest sources of water power.

Label the cities, the bordering country and the bordering states of Oregon and Idaho.

willow goldfinch

MORE TO LEARN

Mark each statement **(T) TRUE** or **(F) FALSE**. On another sheet of paper, rewrite the FALSE statements so they are TRUE.

___ All the land in Washington looks the same.

___ Mt. Rainier is the highest point.

___ The Grand Coulee Dam is the nation's largest concrete dam.

___ Olympia is the capital.

___ In the east are high mountains and thick forests.

___ Water is an important natural resource.

___ The Grand Coulee Dam is on the Snake River.

___ The industrial centers are in the western lowlands.

coast rhododendron

CAN YOU FIND . . . What volcanic mountain erupted in 1980?

_____

West Virginia

the Mountain State

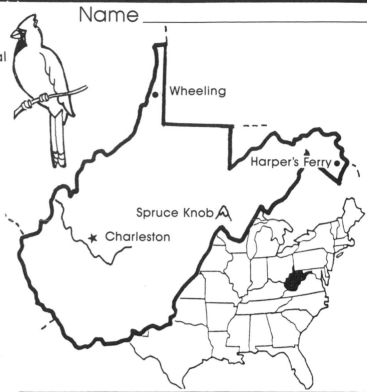

cardinal

Name _____

West Virginia was born during the Civil War — as a result of the Civil War. It became the 35th state in 1863 because it had separated from Virginia when it seceded from the Union.

A rugged mountain land with no level ground and rocky soil, the people have had to look beneath it for a livelihood. Fortunately, there is an abundance of natural resources. — Coal is found under half the state, along with an abundance of natural gas, timber and salt deposits. And, tourism is developing as more visitors discover the beautiful mountains, forests, rivers and natural springs.

Label the bordering states starting with Pennsylvania to the north moving clockwise: Maryland, Virginia, Kentucky and Ohio.

MORE TO LEARN

rhododendron

Use the rhyming clues to fill in the blanks.

First state capital, Wh _____ (rhymes with feeling)

Site of John Brown's Raid, H _____ F _____ (rhymes with warpers and merry)

First state governor, Arthur B _____ (rhymes with core and tan)

Disease common among coal miners, bl _____ l _____ (rhymes with sack and sung)

Ranks second to Kentucky in production of c _____ (rhymes with goal)

Highest point in the state, Sp _____ Kn _____ (rhymes with truce and lob)

Write the circled letters. _____

When unscrambled, they spell the name of the current U.S. Senator from a well-known family. _____

CAN YOU FIND . . . What town changed hands 56 times during the Civil War?
_____

Wisconsin

the Badger State

Wisconsin, the 30th state, is known as "America's Dairyland". It is the nation's leading producer of milk and cheese.

But manufacturing is the state's chief industry. It is a leader in manufacturing engines and turbines, in canning vegetables and brews more beer than any other state.

The natural beauty and 8000 plus lakes attract millions of vacationers. Wisconsin has won fame for its high quality of progressive government and education.

wood violet

Label Madison, the capital, the other cities, and bordering states and Great Lakes starting with Lake Superior on the north and moving clockwise: Michigan, Lake Michigan, Illinois, Iowa, Minnesota and Lake Superior.

MORE TO LEARN

Write the number of the name in Column **B** that matches the phrase in Column **A**. Think logically!

A	B
___ A bordering Great Lake	1. Joseph McCarthy
___ Celebrated pro football team	2. Winnebago
___ Most important type of farming	3. Green Bay Packers
___ Brothers who started their circus there	4. Milwaukee
___ Political party founded at Ripon	5. Dairying
___ Controversial U.S. Senator elected in 1946	
___ The largest city (located on Lake Michigan)	6. Superior
	7. Ringling
___ State's largest lake	8. Republican

robin

CAN YOU FIND . . . Who was the famous architect from Wisconsin?

_____

49

Wyoming

the Equality State

Name _____

Indian paint brush

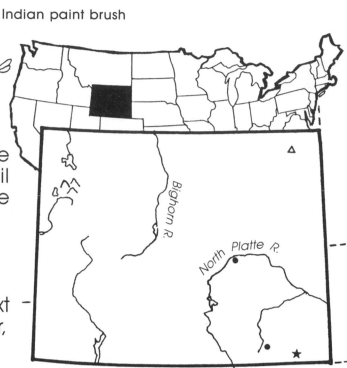

Wyoming, the 44th state, is known for its beautiful mountains and parks. Millions of tourists visit the state each year.

The soil is poor, so 95 percent of the land is used for grazing. Cattle and oil are the foundation of the economy. Since half the land is federally-owned and controlled, the government plays an important role in Wyoming's future.

Cheyenne, the capital and largest city, has less than 50,000 people. The next largest cities are Laramie and Casper, on the North Platte River.

Label the capital, cities, national monument and parks, and bordering states starting with Montana on the north, moving clockwise: South Dakota, Nebraska, Colorado, Utah and Idaho.

MORE TO LEARN

Fill in the blanks using the **WORD BANK**.

WORD BANK	Devils Tower Shoshone	Grand Tetons Esther Morris	Yellowstone Sacajawea

The first national monument _ _ _ _ _ _ _ _ _ _ — ◯

The first, and largest, national park _ _ _ _ _ — ◯ _ _ _ _ _ _

The first national forest _ _ _ _ _ _ _ — ◯

meadow lark

The nation's first woman justice of the peace

_ _ _ _ _ _ _ _ ◯ _ _ _

The Shoshone girl guide of Lewis and Clark _ ◯ _ _ _ _ _ _ _

One of the world's most dramatic mountains _ _ _ ◯ _ — _ _ _ _ _ _

Write the circled letters. _____ Unscramble them to find the name of the site of the first U.S. long range missile squadron.

_____ Air Force Base

CAN YOU FIND . . . What famous family helped the Grand Tetons become a national park? _____

The Grand Canyon

The Grand Canyon in Arizona was once covered by water. The earth's crust began to move over two billion years ago and raised itself one and a half miles above sea level. This mountain-building process and rushing water began to carve the canyon. The Colorado River started shaping the canyon about six billion years ago. As the river cut its way down, it exposed several different kinds of rock which make up the canyon's walls. The canyon is about one mile deep and is anywhere from 6 feet to 18 miles wide. The Colorado River flows through it for 277 miles. Visitors may view the canyon from several points along the North and South Rims, or they may hike down on one of its many trails.

Besides its geologic history, the Grand Canyon has a history of man. Several Indian tribes have lived in it the past 4000 years. Now a small tribe called the Havasupai live at the bottom of the canyon. Spanish explorers were the first to see the canyon. John Wesley Powell, an American geologist, was the first to travel through it. He gave the "big canyon" its name.

Write True or False in front of the statements below.

____ The Grand Canyon began to form two billion years ago.
____ The Grand Canyon is in Colorado.
____ Spanish explorers named the Grand Canyon.
____ The canyon is 18 miles deep.
____ There was once a sea covering the Grand Canyon.
____ The canyon first began when the Colorado River cut through it.
____ There are two rims around the Grand Canyon.
____ John Wesley Powell was the first to see the Grand Canyon.
____ The Grand Canyon is made of several kinds of rock.
____ No one can live in the canyon.
____ The only way to see the canyon is from the top.
____ The Grand Canyon is named as it is because of its size.
____ The formation of the canyon began when the earth's crust erupted.

51

Niagara Falls

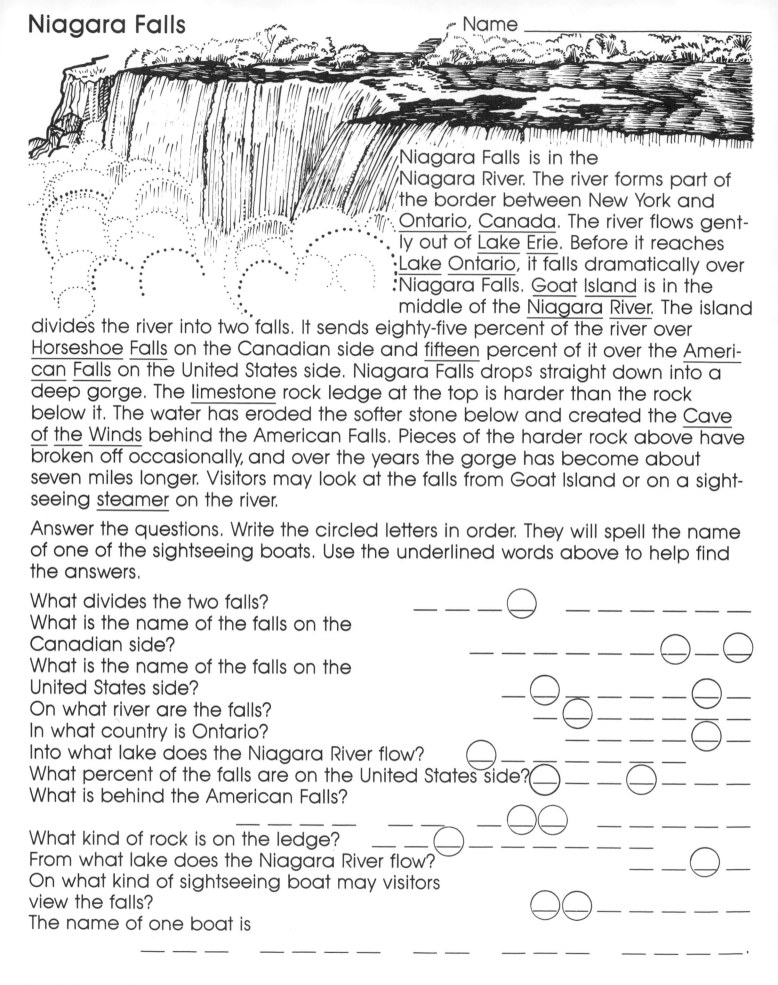

Niagara Falls is in the Niagara River. The river forms part of the border between New York and Ontario, Canada. The river flows gently out of Lake Erie. Before it reaches Lake Ontario, it falls dramatically over Niagara Falls. Goat Island is in the middle of the Niagara River. The island divides the river into two falls. It sends eighty-five percent of the river over Horseshoe Falls on the Canadian side and fifteen percent of it over the American Falls on the United States side. Niagara Falls drops straight down into a deep gorge. The limestone rock ledge at the top is harder than the rock below it. The water has eroded the softer stone below and created the Cave of the Winds behind the American Falls. Pieces of the harder rock above have broken off occasionally, and over the years the gorge has become about seven miles longer. Visitors may look at the falls from Goat Island or on a sightseeing steamer on the river.

Answer the questions. Write the circled letters in order. They will spell the name of one of the sightseeing boats. Use the underlined words above to help find the answers.

What divides the two falls?
What is the name of the falls on the
Canadian side?
What is the name of the falls on the
United States side?
On what river are the falls?
In what country is Ontario?
Into what lake does the Niagara River flow?
What percent of the falls are on the United States side?
What is behind the American Falls?

What kind of rock is on the ledge?
From what lake does the Niagara River flow?
On what kind of sightseeing boat may visitors
view the falls?
The name of one boat is

___ ___ ___ ___ ___ ___ ___ ___

The Great Plains

Name _____

The Great Plains extend 2500 miles north and south from Canada to Texas, and east and west from the Rocky Mountains 400 miles to the central lowlands. Texas, Oklahoma, Kansas, Nebraska, South Dakota, North Dakota, New Mexico, Colorado, Wyoming and Montana are part of the Great Plains. The western boundary is higher and slopes toward the eastern boundary at ten feet per mile. It is not noticeable when one looks out into the distance.

Pioneers crossed the Great Plains in wagons. It was a long trip with little to see but grasslands. It is much the same way today, but the highway is improved. There is still grassland where wheat does not grow on the farms that have been built. There are no bison or buffalo.

Fill in the boxes below with the names of the states that are part of the Great Plains. One letter has been put in to get you started. When you have written the states' names in the boxes, write them again alphabetically on the lines to the right.

The Colorado Plateau

Name _____

Utah | Colorado

Arches Nat'l. Park

Capitol Reef Nat'l. Park

Bryce Nat'l. Park

Cedar Breaks

Colorado Nat'l. Mon.

Canyonlands Nat'l. Park

Glen Canyon

Natural Bridges Nat'l. Mon.

Zion Nat'l. Park

Lake Powell

Rainbow Bridge Nat'l. Mon.

Grand Canyon

New Mexico

Arizona

The Colorado Plateau is a little larger than what is called the Four Corners Region. The area of the Colorado Plateau covers the northern third of Arizona, southern two-thirds of Utah, northwestern quarter of New Mexico and the southwestern quarter of Colorado. The plateau is flat with deep cut canyons. The Grand Canyon is the best known. The Colorado River cuts through several of the canyons; the Colorado National Monument, past the southern end of Arches National Park, Canyonlands National Park, Glen Canyon and through the Grand Canyon. Lake Powell was created when Glen Canyon Dam was built to hold back the Colorado River. Rainbow Bridge National Monument may be reached by boat on Lake Powell. The Colorado River was not the main force to shape some of the places in the region. Natural Bridges National Monument was cut by two other rivers. Zion National Park was cut by the Virgin River. Cedar Breaks National Monument, Capitol Reef National Park and most of Arches National Park were created by rain, streams, wind and ice.

Alphabetize the names of the underlined places. If lined up correctly, each will form a sentence about itself.

_____ has a formation called Landscape Arch.

_____ has fourteen valleys filled with imaginative shapes like monks and temples.

_____ is cut by the Green River.

_____ has a white-capped sandstone ridge.

_____ has steep, pink, cliff formations.

_____ has a shape called Devil's Kitchen.

_____ is a national recreation area.

_____ has one rim higher than the other.

_____ was created when Glen Canyon Dam was built.

_____ has three natural sandstone bridges.

_____ is the world's largest known natural bridge.

_____ has a rock mass called Checkerboard Mesa.

Deserts

There are other desert regions in the southwestern United States besides Death Valley and the Great Basin. The Sonoran desert in southwestern Arizona and southeastern California is not a deserted, sandy place as one might think. All sorts of animal and plant life live in it. Cactus is the main plant. Its flower provides food for birds. Woodpeckers make their homes in the cacti. Lizards and other ground animals often sleep during the day and hunt for their food at night. The Colorado, Yuma and Succulent Deserts are part of the Sonoran Desert. The Mojave Desert is in California between the Sierra Nevada mountains and the Colorado River. The Mojave Desert was once covered by the Pacific Ocean, but when the mountains rose they kept out the water from the sea. The Painted Desert is in northern Arizona. Buttes, mesas, pinnacles, valley formations and varied colors cover the area.

Circle the names of eight American deserts in the puzzle. Write them in alphabetical order on the lines to the right. The names read ⇄ ↓↑ ╱╱.

```
P A I N T E D D E S E R T N
A S E R T S O N O U R N A I
I O U C C U D G V C H O M S
N N H C O L A R O D A O U A
T O V O C G R M V A J O Y B
D R A L C U O E T A M U J T
E A L O U R L T V L A O D A
S N L R L G O E A T B A S E
E A E A E N C T N A S I B R
R U Y E L L A V H T A E D G
```

_____

_____

_____

_____

_____

_____

_____

_____

55

Indian Ruins

The Four Corners region in the United States, where Arizona, Utah, Colorado and New Mexico meet, was occupied hundreds of years ago by the ancestors of today's American Indians. The remains of their dwellings, pottery and tools tell what their life was like about 800 years ago. The United States Government has set many of these sites aside as national monuments and parks to preserve their history and scenic beauty.

Names of some of these sites are scrambled below in a sentence about them. Unscramble the name and write it under the picture of its ruin.

<u>N</u> <u>A</u> <u>M</u> <u>E</u> <u>U</u> <u>O</u> <u>M</u> <u>Z</u> <u>T</u>   <u>S</u> <u>C</u> <u>E</u> <u>T</u> <u>L</u> <u>A</u> National Monument
3 9 1 5 7 2 8 6 4   12 10 15 13 14 11 is a five story building built in a niche on the side of a cliff.

<u>E</u> <u>S</u> <u>M</u> <u>A</u>   <u>R</u> <u>V</u> <u>E</u> <u>E</u> <u>D</u> National Park has many structures that were built beneath an overhang in a cliff.
2 3 1 4   7 5 6 9 8

<u>L</u> <u>E</u> <u>N</u> <u>R</u> <u>B</u> <u>I</u> <u>D</u> <u>A</u> <u>E</u> National Monument has the ruins of a meeting house that looks something like a maze. This was done for defense.
6 5 3 9 1 7 4 2 8

<u>V</u> <u>N</u> <u>O</u> <u>A</u> <u>A</u> <u>J</u> National Monument has three ancient cliff dwellings and sits in the middle of the Navajo Indian Reservation today.
3 1 6 2 4 5

_ _ _ _ _ _ _ _ _

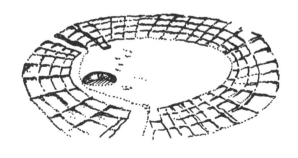

_ _ _ _ _ _ _ _ _   _ _ _ _ _ _

56

The Mississippi River

Name _____

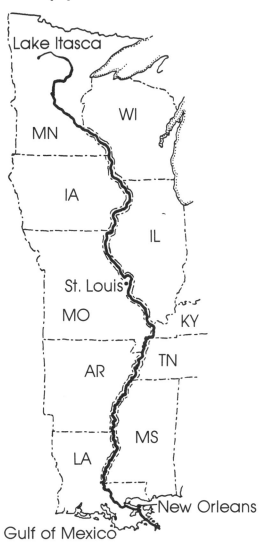

Lake Itasca

WI
MN
IA
IL
St. Louis
MO
KY
TN
AR
MS
LA
New Orleans
Gulf of Mexico

Chippewa Indians gave the name Messipi to the Mississippi River which means great river. The Mississippi is the longest river in the United States. It is 2348 miles long. Its source is a stream flowing out of Lake Itaska in northwest Minnesota. Its beginning is so small you can wade across it. As it travels, thousands of streams and rivers join it to make it the large river it is. The Mississippi River forms all of part of the eastern borders for the states of Minnesota, Iowa, Missouri, Arkansas and Louisiana and the western borders for Wisconsin, Illinois, Kentucky, Tennessee and Mississippi. As the river travels, it moves faster and picks up and carries sand, mud and pebbles. This is deposited as silt along the lower banks and in the mouth of the river. The river ends its journey in the Gulf of Mexico.

There is a rhythm to spelling M-I-S-S-I-S-S-I-P-P-I that is fun. The answer to each question below begins with a letter of the river's name. Let the underlined words above help you.

In what state does the Mississippi
River begin?   M _ _ _ _ _ _ _ _ _

What is one state it borders?   I _ _ _ _

What is something it carries?   S _ _ _ _

What is something it deposits?   S _ _ _

What is the lake's name where it begins?   I _ _ _ _ _ _

What does the river begin as?   S _ _ _ _ _ _

What is another word that means beginning?   S _ _ _ _ _ _

Who named it Messipi?   I _ _ _ _ _ _ _

What does it do with mud and sand?   P _ _ _ _ _ _ _

What else does it pick up?   P _ _ _ _ _ _ _

What is another state it borders?   I _ _ _ _ _ _ _ _

Tributaries of the Mississippi

Name _____

A tributary is a smaller stream or river that empties into a larger body of water. Every stream or river that flows must go somewhere. Almost half of the rivers in the United States empty into the Mississippi River. These rivers flow between the Rocky Mountains in the west and the Appalachian Mountains in the east. The Missouri is the second longest river. It flows into the Mississippi just above St. Louis, Missouri, after traveling 2,315 miles from its source in Montana, through South Dakota, Nebraska, Iowa, Kansas and Missouri. It was nicknamed Big Muddy by the pioneers because it was so dirty. The Arkansas River begins high in the Rockies in Colorado and flows through Kansas, Oklahoma and Arkansas before it empties into the Mississippi River. The Ohio River flows 981 miles through coal fields, steel factory districts and farm lands. It flows from Pennsylvania along the borders of West Virginia, Kentucky, Ohio, Indiana and Illinois. It empties into the Mississippi at its widest point—Cairo, Illinois.

Write True or False to the following statements.

_____ The Missouri River is longer than the Mississippi.

_____ The Arkansas River begins in Colorado.

_____ The Ohio River is the smallest of the rivers mentioned.

_____ The Mississippi is a tributary to these rivers.

_____ The Ohio empties into the Mississippi just above St. Louis.

_____ Pennsylvania is where the Missouri River begins.

_____ The Missouri River goes through or by six states.

_____ Over half the rivers from the Rockies to the Appalachians flow into the Mississippi.

_____ The Ohio River flows through Oklahoma, Kansas and Arkansas.

_____ The Missouri's widest point is in Cairo, Illinois.

_____ Big Muddy is a good name for the Arkansas River.

_____ The Arkansas River is larger than the Ohio River.

_____ The Appalachian Mountains are in the east.

_____ Sometimes a river can be a border between states.

Rivers in the West

There are several important rivers in the western United States that do not empty into the Mississippi River. They are best known for their beauty and recreational activities including rafting, water skiing and fishing. There is not as much water in the western United States so the rivers there have also been used in the building of dams, lakes and reservoirs.

Work the math problems below. The answers to the problems will spell the name of some of the rivers in the west. Use the CODE BOX on the left to decode each river's name. Write it out correctly and read a fact about that river.

CODE BOX

A = 26	N = 25
B = 24	O = 23
C = 22	P = 21
D = 20	Q = 19
E = 18	R = 17
F = 16	S = 15
G = 14	T = 13
H = 12	U = 11
I = 10	V = 9
J = 8	W = 7
K = 6	X = 5
L = 4	Y = 3
M = 2	Z = 1

$$\begin{array}{cccccccc} 30 & 32 & 16 & 5 & 11 & 13 & 28 & 17 \\ -8 & -9 & -12 & +6 & -9 & +11 & -18 & +9 \end{array}$$

___ ___ ___ ___ ___ ___ ___ ___

This river is the second longest river in the Western Hemisphere to empty into the Pacific Ocean.

$$\begin{array}{ccccc} 7 & 12 & 13 & 19 & 25 \\ +8 & +13 & +13 & -13 & -7 \end{array}$$

___ ___ ___ ___ ___

This river is a tributary of the Columbia River.

$$\begin{array}{cccccccc} 15 & 42 & 11 & 14 & 9 & 15 & 14 & 17 \\ +7 & -19 & -7 & +9 & +8 & +11 & +6 & +6 \end{array}$$

___ ___ ___ ___ ___ ___ ___ ___

This river has carved the Grand Canyon for millions of years.

$$\begin{array}{ccccccccc} 23 & 3 & 12 & 7 & 28 & 19 & 14 & 36 & 29 \\ -6 & +7 & +11 & +7 & -11 & +7 & +11 & -16 & -11 \end{array}$$

___ ___ ___ ___ ___ ___ ___ ___ ___

This river forms an international border between the United States and Mexico.

The Great Lakes

Name _____

The Great Lakes began to form over 250,000 years ago when large sheets of ice, called glaciers, dug holes in the land. Melting glacial snows filled the holes with water and formed the Great Lakes, the world's largest group of freshwater lakes. There are five lakes (in order of size): Superior, Huron, Michigan, Erie and Ontario. It is possible to travel on a boat from the Atlantic Ocean, through the Saint Lawrence Seaway, over Lakes Ontario, Erie, Huron and Michigan, onto the Illinois River, into the Mississippi River to the Gulf of Mexico.

Use the map and information above to help answer the questions below. Write the circled letters in your answers on the blanks at the bottom of the page. They will spell the name of the shallowest Great Lake.

Color Lake Superior blue, Lake Michigan green, Lake Huron orange, Lake Erie red and Lake Ontario purple.

Which lake is the largest? __ __ __ __ __ __ __ __

Which lake does not have Canada as one of its boundaries?

__ __ __ __ __ __ __ __

How many lakes form the Great Lakes? __ __ __◯

From what did the water come to fill the Great Lakes?

__ __ __ __ __ __ __ __

Which lakes touch New York? __ __ __ __ and __ __ __ __◯ __ __

Which state is touched by Lakes Michigan and Superior?

__◯__ __ __ __ __ __ __

What Michigan city is an important port on the Great Lake system?

__◯__ __ __ __ __

The shallowest Great Lake is Lake __ __ __ __ .

The Great Salt Lake and Its Surroundings

The Great Salt Lake and Great Salt Lake Desert are part of the Great Basin. The Great Basin is a large desert area extending into parts of six states; California, Idaho, Oregon, Nevada, Utah and Wyoming. It is called a basin because all of its bodies of water stay within it. Sinks lie in some of its valleys. The Great Salt Lake is its largest sink.

The Great Salt Lake is in Utah. It is considered one of the natural wonders of the world. A freshwater lake was once in its place thousands of years ago, but it dried up leaving several small lakes. The Great Salt Lake is the largest of the remaining lakes. Today the lake gets its water from rain and freshwater streams, but its salt is from the salt deposits left by the original dried-up lake. When the lake is low, it has several islands. Cattle are raised on the largest one, Antelope Island. The smaller islands are breeding grounds for gulls, ducks, geese and pelicans. There are no fish in the lake, only shrimp. The Great Salt Lake Desert is a low, flat, dry region to the west of the lake. A part of it is so hard that it is used as a track for automobile racing.

Use the underlined words above to help answer the clues below. Write your answers on the blanks. The name geologists give the Great Salt Lake will appear down in the boxes.

Place in which water stays   ☐ _ _ _ _

State into which Great Basin extends   _ _ _ _ ☐

Island where cattle raised   _ ☐ _ _ _ _ _ _

What is in some valleys   _ _ ☐ _ _

Birds that nest on smaller islands   _ _ _ ☐ _

State into which Great Basin extends   _ _ _ ☐ _

Lives in Great Salt Lake   _ _ _ ☐ _ _

Birds that nest on smaller islands   _ _ ☐ _ _

State into which Great Basin extends   _ _ _ _ ☐ _

Birds that nest on smaller islands   _ _ ☐ _ _ _ _

Geologists call it _ _ _ _ _ _ _ _ _ _ _ Lake.

The Appalachian Mountains

Name _____

The Appalachian Mountains extend 1500 miles across the <u>Eastern</u> <u>United</u> <u>States</u> between Alabama and Canada. The <u>Appalachian</u> Mountains are the oldest mountain system in North America and the second largest system after the Rocky Mountains. All the main ranges east of the Mississippi River are in the Appalachians except for the <u>Adirondacks</u>. The ranges in the United States that are a part of this system are the White Mountains, Green Mountains, Catskill Mountains, Great Smoky Mountains, Allegheny Mountains, Cumberland Mountains and the Blue Ridge Mountains. <u>Mount</u> <u>Washington</u> is the tallest mountain in New Hampshire and has a weather station on top. Mount Mitchell, in the <u>Blue</u> <u>Ridge</u> Mountains, is the tallest mountain in the Appalachians. <u>Spruce</u> <u>Knob</u> is the tallest peak in the Allegheny range.

Write the scrambled answer to each clue below on the line following it. Let the underlined words above help you.

Where the Appalachian Mountains are located

ARSNEET _____

NIUETD _____

TTSSAE _____

Mountain range that is not part of the Appalachian system

DIAASRCODNK _____

The oldest mountain system in North America

ACAPAINAPLH _____

Where there is a weather station

ONMUT _____

NSITAGNWHO _____

The mountains in which the tallest Appalachian peak is located

LEUB DEGRI _____

The tallest peak in the Allegheny range

PCUSRE KOBN _____

Write the mountain ranges in alphabetical order on the back of this paper.

The Rocky Mountains

Name _____

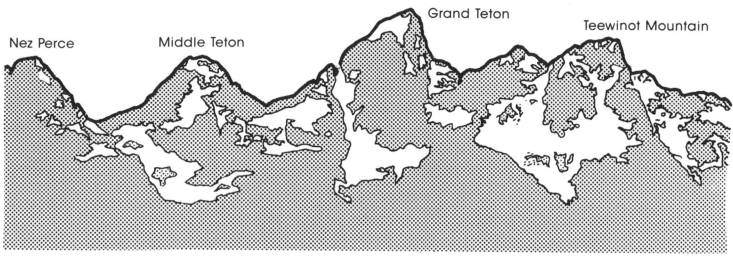

Nez Perce   Middle Teton   Grand Teton   Teewinot Mountain

The Rocky Mountain chain is the largest mountain system in North America. It extends more than 3000 miles through the United States and Canada. It is 350 miles wide in some places. In the United States the range extends through New Mexico, Colorado, Utah, Wyoming, Idaho, Montana and Alaska. The high peaks in the Rockies form the Continental Divide which separates the direction water flows—toward the Atlantic or Pacific Oceans. It crosses New Mexico, Colorado, Wyoming, Idaho and Montana in the United States. The Columbia, Missouri, Colorado, Arkansas and Rio Grande Rivers begin high in the Rockies. Yellowstone, Grand Teton and Rocky Mountain National Parks are in the Rockies as is Waterton/Glacier International Peace Park.

Circle as directed in the puzzle below. Some names will be circled two or three times.

1. The names of the states the Rocky Mountain system is in. Circle them red.

2. The names of the National Parks in the Rocky Mountains. Circle them green.

3. The names of the rivers that begin in the Rockies. Circle them blue.

4. The names of the states the Continental Divide crosses. Circle them yellow.

```
E G R A N D E O I W O L L E Y W
W R A N C H E R I O Y M R N O E
A A A N A T N O M L O O A T M D
I N T D A R O O L O C U M R I N
R D N E R G T M I K G O I I G A
U T A H R I S Y Y E N A S O N R
O E O M A T W M N T I I C E E G
S T T O R H O H A T M B O N W O
S O R N A U L N R M U M L O M I
I N E A N A L A G R Y U A T E R
M I T T A M E C L L W L M S X A
C H A U T E Y W A E A O B O I D
O I D A H O A R K S A C A L C O
N R U S A S N A K R A U I L O H
O O R I O O D A R O L O C E A D
M I S O U R R I C O L U M Y R I
```

The Cascades

The Cascade Range is a chain of mountains that extend from northern California, through western Oregon and Washington, into British Columbia. The range is named after large cascades the Columbia River makes as it cuts through the mountains.

The mountains in this range were made from lava flows occurring over thousands of years. Most of the range's peaks are extinct volcanoes. Mount Rainier and Mount Adams in Washington, Mount Hood in Oregon and Mount Shasta in California are four inactive volcanoes. Crater Lake, in the southern part of the range, is all that remains of ancient Mount Mazama which erupted 7000 years ago. Lassen Peak in California was inactive for many years, but from 1914–1921 it had several eruptions. Mount Saint Helens in Washington had been quiet for over 100 years, when in 1980 it began to erupt again. More than 1000 feet of its peak has been blown off in its recent eruptions, and a large crater has developed. Scientists do not know when the next eruption will be, but they keep an eye on the Cascades.

Answer True or False to the following statements.

_____ The Cascades were made from volcanic eruptions over thousands of years.

_____ Most of the volcanoes in the Cascades are extinct.

_____ Mount Saint Helens had never erupted before 1980.

_____ Lassen Peak is in Oregon.

_____ The Cascade Range extends from Northern California into Canada.

_____ Crater Lake is a volcano.

_____ Mount Saint Helens has become higher since 1980.

_____ The Columbia River cuts through the Cascades.

_____ Mount Rainier and Mount Hood are inactive volcanoes.

_____ Lassen Peak has not erupted in over 100 years.

_____ The Cascades never have any volcanic activity now.

_____ Scientists are interested in the Cascades.

_____ Mount Rainier and Mount Adams are in Washington.

_____ From 1912–1941 Lassen Peak had several eruptions.

_____ When a volcano is active it erupts.

64

Coast Ranges

The Coast Ranges are a chain of mountains extending from Alaska to Mexico. They include several ranges. To the north they are a continuation of the Sierra Nevada and Cascade Ranges. They continue on as the Alaska and Aleutian Ranges. The Sierra Nevada extend in a north–south direction for 400 miles in California. Mount Whitney, the highest peak in the 48 states, is in the Sierra Nevada. Yosemite, Sequoia and Kings Canyon National Parks are part of the Sierra Nevada. Directly north are the Cascades. There are over 100 glaciers in the Olympic Mountains in northern Washington, part of the Pacific Coast Range. Mount Olympus is the highest peak in these mountains. The Alaska and Aleutian Mountains end the Coast Ranges. Mount McKinley, in the Alaska Range, is sometimes called "Top of the Continent". It is the highest point in North America.

The correct answers to the clues below will spell COAST RANGES down. Use the underlined words above to help you fill in the blanks.

1. Range Olympic Mountains belong to
2. Highest peak in the Olympic Mountains
3. Range Mount McKinley is in
4. Range just north of the Sierra Nevada
5. Highest peak in the 48 states
6. Range in California
7. National Park in Sierra Nevada
8. Highest Mountain in North America
9. National Park in Sierra Nevada
10. Range after Alaska Range
11. National Park in Sierra Nevada

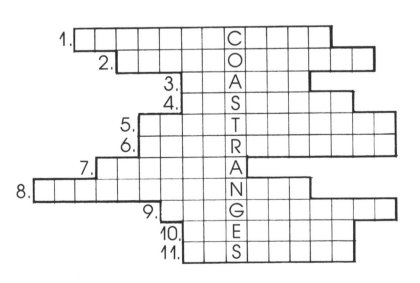

Death Valley National Monument

Name _____

Death Valley is a desert. The lowest land surface in the Western Hemisphere, 282 feet below sea level, is located in Death Valley. Death Valley's highest point, 11,049 feet, is on top of Telescope Peak in the Panamint Mountains.

Death Valley was named by some pioneers in 1849, who after crossing it, could not see how any life could survive in its dry, hot climate. Usually less than two inches of rain falls in a year. Summer temperatures average 125°F a day. The hottest temperature ever recorded was 135°F. The lowest temperature ever recorded was 15°F. Winter temperatures usually average 90°F. Some animal and plant life have been able to adapt to the desert conditions. In the 1800's, miners lived and worked in the valley. The towns they lived in are ghost towns now. But their burros survived and run wild along with lizards, snakes, desert bighorn sheep and rodents. One man, Walter Scott, lived in Death Valley for thirty years. The "castle" he built is still there.

Write True or False in front of the statements below.

_____ Some animal and plant life have adapted to Death Valley's climate.
_____ There are mining towns in Death Valley.
_____ The highest point in the Western Hemisphere is located in Death Valley.
_____ Death Valley is a desert.
_____ Death Valley was named by some miners in 1849.
_____ Living is easy in Death Valley.
_____ The highest point in Death Valley is in the Panamint Mountains.
_____ It is usual for Death Valley to receive less than two inches of rain in a year.
_____ Visitors to Death Valley will see no signs of life.
_____ Normal summer temperatures are 135°F.
_____ The climate in Death Valley is hot and dry.
_____ Winter temperatures average 90°F.

Badlands National Park

Rain, wind and frost have carved ravines, ridges, low hills and cliffs in the South Dakota prairie. These same weather conditions have also exposed the rock so the layers that were laid down millions of years ago are clearly visible. Prehistoric animal and swamp plant fossils are part of the layers, which means the Badlands were once warm and moist. Since the land has been known to white man, it has been dry and bare. The temperatures are very hot in summer and cold in winter. The Badlands were named because they were difficult for the settlers to cross and impossible to farm. Some wildlife has been able to adapt to the Badlands. Unscramble their names below to see what some of the wildlife looks like today in Badlands National Park.

C U Y A C _____
3 2 1 5 4

N O W O D O T O T C
6 2 7 5 10 9 4 8 3 1

Y T C O O E _____
3 5 1 4 2 6

D R E G A B _____
3 6 5 4 2 1

G D O N L E   A L E G E
1 4 2 6 3 5   8 10 11 9 7

S A T E R T K L E N A
7 2 4 11 1 3 10 5 6 8 9

Everglades National Park

Everglades National Park is a subtropical region in southern Florida. It is the third largest national park in the United States and one of the country's wettest. The Park includes Ten Thousand Islands on the Gulf of Mexico and the Big Cypress Swamp. The park is a small part of the Florida Everglades. Half of the park is under seawater. Raised islands, called hammocks, are scattered in the marshy sawgrass. The park as a refuge provides protection for many plants, animals and birds.

Discover some of the Everglades wildlife. Label each picture with one of the four titles to the right. Color the finished pictures.

ALLIGATOR AND MANATEE
EGRET
ORCHID
MANGROVE AND CYPRESS TREES

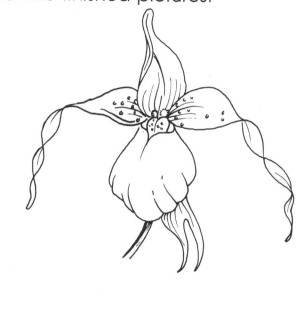

_ _ _ _ _ _ _

_ _ _ _ _ _ _ _ _ _ _

_ _ _ _ _ _ _ _ _ _

_ _ _ _ _ _ _

Grand Teton National Park

Name _____

Grand Teton National Park is an area of mountains, lakes and forests in northwestern Wyoming just six miles south of Yellowstone National Park. The two parks are connected by the John D. Rockefeller Memorial Parkway. Grand Teton National Park was established in 1929 after it was unable to become a part of Yellowstone. In 1950 Congress enlarged Grand Teton by making Jackson Hole National Monument a part of it.

The Grand Teton mountains are the youngest of the Rocky Mountain system. About nine million years ago, part of the earth's crust moved upward to form the mountains, and part moved downward to form the valley floor (Jackson Hole). Seven of the Teton peaks rise 12,000 feet above sea level. The Grand Teton is the highest. Several mountain lakes are at the foot of these steep mountains on the valley floor. Jackson Lake is the largest. The Snake River flows out of it.

The names of five other glacial lakes are below. Unscramble them and write their names on the blank lines.

N E N Y J _____ Lake
3 2 4 5 1

G T A R T G A _____ Lake
4 1 5 6 7 3 2

P L E H P S _____ Lake
5 4 3 2 1 6

G L H E I _____ Lake
4 1 5 2 3

L R E B Y A D _____ Lake
5 2 6 1 7 3 4

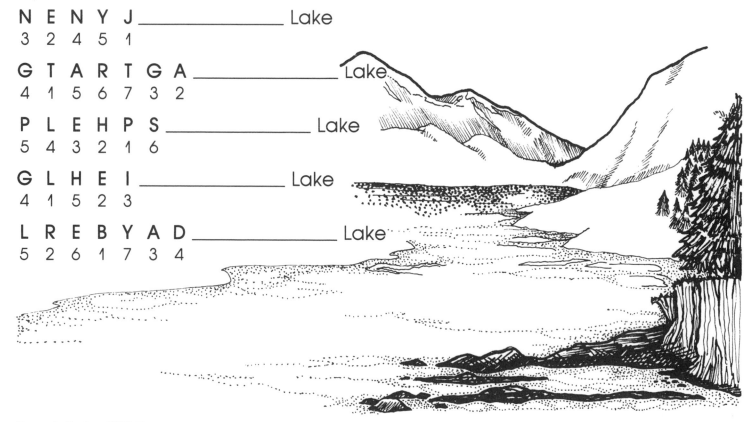

Great Smoky Mountains National Park

Great Smoky Mountains National Park is on the boundary between North Carolina and Tennessee. It is part of the Appalachian mountain system. Almost all of the Smoky mountains lie within the park. Great Smoky Mountains National Park has sixteen peaks above 6000 feet. Clingman's Dome is the highest. There are over 150 different kinds of trees, and hundreds of shrubs and flowering plants, including several kinds of orchids. Indians called these mountains SHA-GONIGEI which means "little blue smoke". The thick vegetation that covers the park combines with the water to produce a water vapor that covers the park with a smoky mist or haze. That is how the mountain range was named.

Below are some names of some of the mountain peaks over 6000 feet. Unscramble their names and write them on the blanks. The circled letters will spell out the name of the area's first settlers.

S O L N C I L L
7 2 3 6 1 5 4

C P M H N A A
1 4 5 2 7 6 3

T M   E L   O T R C E
2 1   4 3   6 8 7 5 9

Y O G U T
3 4 1 2 5

P T K E R A H
3 7 1 2 6 5 4

G B I   E A L E C O H T C A O   B O N K
3 1 2   14 5 8 13 4 9 12 6 11 7 10   18 17 16 15

Petrified Forest National Park

Name _____

Petrified Forest National Park in Arizona was once a flat floodplain with streams wandering over it. Large trees grew to its south. Reptile-like animals lived in the area too. The trees fell and were washed onto the floodplain. The land sank. The trees and other wildlife were buried under mud, silt and ash. Because no air could reach the trees, they changed from wood to rock—called petrified wood. Millions of years later when the earth moved, it raised the trees above ground. The fallen trees were exposed as petrified wood. Fossils of other ancient plants and animals were also raised. That is how we know today what the area looked like 225 million years ago. When the park was "discovered", souvenir hunters took the wood. In order for it not to disappear, six forests were set aside, and a national park was established.

Work the math problems below. The answers will spell out what scientists discovered recently in the park. Use the code box on the left to decode the message. Write it out under the problems. Read the answer down in columns.

CODE BOX

A = 13	N = 26
B = 12	O = 25
C = 11	P = 24
D = 10	Q = 23
E = 9	R = 22
F = 8	S = 21
G = 7	T = 20
H = 6	U = 19
I = 5	V = 18
J = 4	W = 17
K = 3	X = 16
L = 2	Y = 15
M = 1	Z = 14

$5 + 5$ = ___ ___

$22 - 17$ = ___ ___

$14 + 12$ = ___ ___

$17 + 8$ = ___ ___

$28 - 7$ = ___ ___

$7 + 6$ = ___ ___

$25 - 6$ = ___ ___

$17 + 5$ = ___ ___

$9 + 12$ = ___ ___

$8 - 5$ = ___ ___

$18 - 9$ = ___ ___

$11 - 9$ = ___ ___

$6 + 3$ = ___ ___

$12 + 8$ = ___ ___

$13 + 12$ = ___ ___

$19 + 7$ = ___ ___

A __ __ __ __ __ __ __   __ __ __ __ __ __ __ nicknamed Gertie

Name _____

Sequoia, Kings Canyon and Redwood National Parks

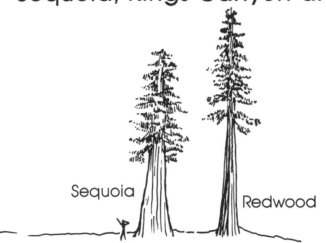

Sequoia

Redwood

Sequoia, Kings Canyon and Redwood National Parks are in California. Sequoia and Kings Canyon are next to each other on the western slopes of the Sierra Nevada range in the central part of the state. Redwood National Park is along the Pacific coast in northern California and southern Oregon. Sequoia and Kings Canyon are two separate parks, but are run as one by the National Park Service. They contain some of the highest peaks of the Sierra Nevada and some of the oldest and largest trees in the world. Mount Whitney, the tallest mountain in the 48 states, is in Sequoia National Park. General Sherman tree in Sequoia has the largest amount of wood, making it the largest tree in the world—272.4 feet high and 101.6 feet around. Redwood National Park has the world's tallest tree—368 feet tall.

Write the answers to the questions below on the blanks. When you have answered all the questions, some letters will have a number under them. Write that letter above its number below to learn something about the difference between redwood and sequoia trees.

In what park is the
world's largest tree? __ __ __ __ __ __ __
                        10   5   8

What is its
name? __ __ __ __ __ __ __   __ __ __ __ __ __ __ __
                    9   1

In what mountain range are
Kings Canyon and Sequoia
National Parks
located? __ __ __ __ __ __   __ __ __ __
          7   2   3

In what state are these parks? __ __ __ __ __ __ __ __ __ __
                                          13   12

What is the name of the tallest
peak in the 48 states? __ __ __ __ __ __   __ __ __ __ __
                          11   4   6

__ __ __ __ __ __ __   __ __ __ __ __   __ __ __
 1 2 3 4 5 5 3   6 1 2 2 7   8 1 2

__ __ __ __ __ __ .
 6 8 9 9 2 1

__ __ __ __ __ __ __ __   __ __ __   __ __ __ __ __ __ __ .
 7 2 10 11 5 12 8 7   8 1 2   13 8 6 6 2 1

Yellowstone National Park

Name _____

Old Faithful

Yellowstone National Park is mainly in the northwest corner of <u>Wyoming</u>, but it also extends into <u>Idaho</u> and <u>Montana</u>. It is famous for many natural wonders, but most of all for its more than 200 <u>geysers</u> and thousands of <u>hot springs</u>. Beginning about 2 million years ago, volcanic eruptions occurred here. What is now the park's central portion collapsed forming a <u>basin</u>. The heat under the earth that caused the ancient eruptions is still responsible for the geysers and hot springs that are in the park today. <u>Old Faithful</u> is the most famous geyser. It erupts boiling water about every 65 minutes. <u>Steamboat Geyser</u> is the world's largest. It spouted 400 feet in the air for a record. <u>Grand Prismatic Spring</u> is the largest spring in Yellowstone. <u>Fountain Paint Pots</u> is a series of hot springs and bubbling pools formed by steam and gases rising from holes in the ground. At <u>Mammoth Hot Springs</u>, flowing water deposits limestone that builds terraces. <u>Norris Geyser Basin</u> has hundreds of geysers and hot spring pools. It is the hottest and most active thermal area in Yellowstone. Just north of Yellowstone Lake, the earth's surface is rising a little less than an inch each year. This suggests there will be future volcanic activity.

Answer the questions below. Let the underlined words above help.

In what state is most of Yellowstone National Park located? _____
Into what other states does it extend? _____
For what is Yellowstone best known? _____ and _____
Name the most famous geyser. _____
What is a series of hot springs? _____
Where in the park are hundreds of geysers and hot springs? _____
Which spring is the largest? _____
What was formed after the volcanic eruptions? _____
Which geyser is the largest? _____
Where are terraces formed? _____

Terraces at Mammoth Hot Springs

Name _____

Wildlife In Yellowstone National Park

Yellowstone is the largest wildlife preserve in the United States. Over forty kinds of animals and two hundred birds may be seen there. Look for a bison, elk, grizzly bear, mule deer, trumpeter swan, bighorn sheep, coyote, osprey and pelican among the lodge pole pine in the hidden picture below. Each animal is numbered. Write its name next to its number under the picture.

1. _____   4. _____   7. _____

2. _____   5. _____   8. _____

3. _____   6. _____   9. _____

Yosemite National Park

Name _____

Yosemite National Park is in California in the Sierra Nevada mountains. Five hundred million years ago, it was all under water. Eventually the earth began to twist and turn. Mountains rose out of the sea. The Merced River cut deep into the rock carving a V-shaped valley. But glaciers moved back and forth making it U-shaped. The glaciers cut off many valleys being shaped by smaller streams and caused the formation of "hanging valleys" from which many of Yosemite's waterfalls descend. Yosemite Falls has upper and lower falls. There is a rainbow at the base of Vernal Falls. Nevada Falls was named "Yo-wipe" by the Indians because it means twisted falls in Indian. Bridalveil Falls looks like a bride's headdress. Illilouette Falls is one of the smaller falls descending only 370 feet. Ribbon, Sentinel and Silver Strand Falls usually run only in the spring when there is a lot of moisture. Waterwheel Falls is a series of pinwheels of water in the Hetch Hetchy Valley of the park.

Circle the names of the waterfalls in the puzzle. The names read ⇆ ↑↓ ↗↙ Write their names below the puzzle in alphabetical order.

```
T E S E L E N I T N E S R B N
I T W N S Y O S E M I T E R I
M L I A I R B B O L I A S I L
E T L E T N B N V E R V T D L
S I V I V E I E S E E E R A I
O L E O N R R L N E R N A L L
Y O R U E S A W I N I N N V O
H W R E T A W A H O B B A A U
E L A R I T N E S E B I D L E
E R A D A V E N I T E N T E T
L N B R I D A L V E I L S N T
D Y O S E M E T E N E V A D E
```

1. _____   4. _____   7. _____

2. _____   5. _____   8. _____

3. _____   6. _____   9. _____

More About Yosemite

Name _____

Yosemite National Park has over 760,000 acres of land with 700 miles of trails. From the Valley Visitor Center in the middle of the park, one can set out to explore many of the park's sights. Follow the directions and mark the map below.

With orange, trace over the road from the east entrance, over Tioga Pass, California's highest automobile pass. Stop at the Tuolumne Meadows Visitor Center and go on to the Valley Visitor Center.

With red, trace over the road from the Valley Visitor Center to Glacier Point where you can get a good view of Yosemite. Circle some of Yosemite's famous landmarks: three rock masses; Cloud's Rest, El Capitan and Half Dome; two waterfalls; Bridalveil and Yosemite; and Mirror Lake.

With blue, trace over the road from the Valley Visitor Center to the Pioneer Yosemite History Center to see some historic buildings and horse drawn carriages.

With green, trace over the road from the museum to Mariposa Grove to see the oldest Sequoia tree, the "Grizzly Giant".

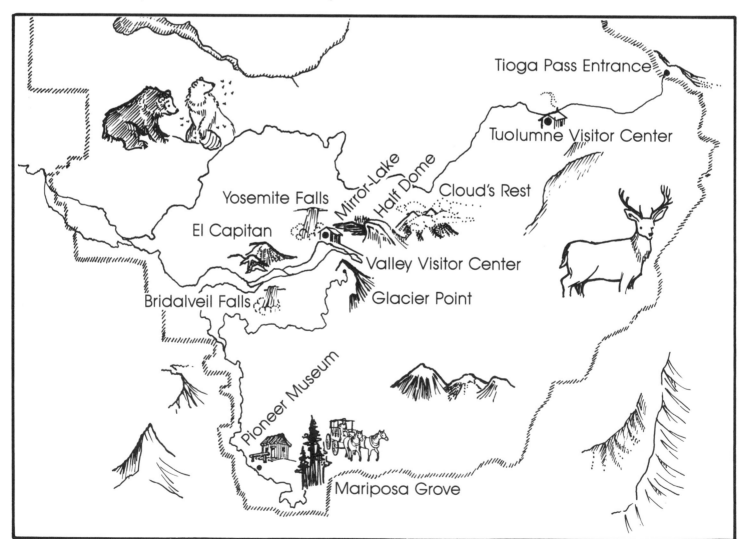

Do You Know These Symbols?

Name _____

Written in the box below are names of some symbols of our heritage. Below the box are pictures of these symbols. Write the name for each one on the line under its picture. Color each symbol as you are directed.

Eagle	Washington Monument	Flag	Liberty Bell
Statue of Liberty	United States Capitol		The White House
Mount Rushmore	Jefferson National Expansion Memorial		

Color me silver.

Color me red, white and blue.

Color me brown.

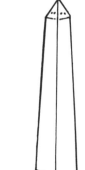

Color me white.

Color me tan.

Color me brown, white and yellow.

Color me green.

Color me gray.

Color me white.

Mount Rushmore

Name _____

Mount Rushmore is a monument of four American President's heads. It is carved into the hard granite Black Hills of South Dakota. The heads stand for the birth and growth of our country. George Washington represents the fight for independence and the start of our nation. Thomas Jefferson stands for the belief in a representative government. Abraham Lincoln represents the unity of the states and equality for all citizens. Theodore Roosevelt stands for the part the United States would take in the world during the twentieth century. The heads were designed by a sculptor named Gutzon Borglum. It took fourteen years to make the monument. Follow the directions below.

Mark a blue **X** on Jefferson's head. Circle Lincoln with black. Draw an orange line under Washington. Color Roosevelt's face yellow.

Solve the crossword.

DOWN
1. President who believed the government should represent the people
2. Sculptor of Mount Rushmore
3. President who kept the states together
4. Mount Rushmore is in South

_____.

ACROSS
5. Number of years it took to build Mount Rushmore
6. President that stood for future of the United States
7. Father of our country

The Alamo

Name _____

The Alamo was built as a Catholic mission in 1718. It was not called the Alamo until later. Alamo is a Spanish word for the cottonwood trees that surrounded the mission. The Alamo was used sometimes as a fort. In 1836 it was the scene of a battle between Mexico and 150 Texans who wanted to keep Texas independent. All the Texans were killed, including Davy Crockett. About a month later, behind the battlecry, "Remember the Alamo", General Sam Houston led new forces in a surprise attack on the Mexican troops. He overthrew them. Today the Alamo is a historic structure in the center of San Antonio, Texas.

Find a picture of the Alamo, its flag, Davy Crockett and Sam Houston in the picture below. Color the fort brown, the flag blue and white, Davy Crockett orange and Sam Houston pink.

Stone Mountain

Name _____

Stone Mountain is a state park near Atlanta, Georgia, and contains the largest stone mountain in North America. It is called Stone Mountain! Its highest point is 700 feet above the base. It is made of granite. The park has a lake, beach, golf course, museum, a restored plantation and a skylift that carries 50 people at a time to the top of the mountain.

A sculpture is carved into the side of the mountain. It honors the men who fought courageously in the Civil War. Three artists took 46 years to complete it.

There are three southern leaders on horseback on the sculpture. Their names are scrambled below in the order they appear on the sculpture. Unscramble their names and write them on the lines. Color as directed.

$$\frac{N}{9} \frac{E}{2} \frac{R}{6} \frac{F}{3} \frac{E}{5} \frac{J}{1} \frac{F}{4} \frac{O}{8} \frac{S}{7} \qquad \frac{D}{10} \frac{V}{12} \frac{S}{14} \frac{A}{11} \frac{I}{13}$$

_____   _____

Color him green.

$$\frac{B}{3} \frac{E}{4} \frac{T}{6} \frac{R}{1} \frac{R}{5} \frac{O}{2} \qquad \frac{E.}{7} \qquad \frac{E}{10} \frac{E}{9} \frac{L}{8}$$

_____   __   _____

Color him red.

$$\frac{T}{2} \frac{E}{5} \frac{A}{7} \frac{S}{1} \frac{N}{4} \frac{L}{8} \frac{O}{3} \frac{L}{9} \frac{W}{6} \qquad \frac{A}{11} \frac{S}{14} \frac{J}{10} \frac{K}{13} \frac{N}{16} \frac{C}{12} \frac{O}{15}$$

_____   _____

Color him blue.

Alcatraz

Alcatraz was a prison for the most dangerous <u>criminals</u>. It was built on a <u>twelve</u> acre rock island, one mile off the coast of <u>California</u>. If a prisoner tried to escape, he <u>drowned</u> or was recaptured before reaching the mainland.

　　Alcatraz was a prison as described above from 1934 to 1963. Earlier it had served as a <u>barracks</u> for Civil War <u>soldiers</u> and a military prison. In 1969 a group of Indians took it over.

In 1972 it became part of the Golden Gate National Recreation Area. The <u>prison</u> still stands on Alcatraz Island in San Francisco Bay and may be visited by tourists.

Answer the questions below using the underlined words from above. When they are written in the correct blanks, the letters reading down in the boxes will spell what Alcatraz means in Spanish.

What was Alcatraz?　　　　　　　　　　☐ _ _ _ _ _

What might have happened to an
　　escaping prisoner?　　　　　　_ _ _ _ _ ☐ _

How many acres are on Alcatraz
　　Island?　　　　　　　　　　_ _ _ ☐ _ _

Who lived there during the Civil
　　War?　　　　　　　　　_ _ _ _ _ ☐ _ _

What state is one mile from
　　Alcatraz?　　　　　　　　☐ _ _ _ _ _ _ _ _ _

In what did the soldiers live during
　　the Civil War?　　　　_ _ _ _ _ ☐ _ _

What sort of people lived there
　　from 1934–1963?　　_ _ _ _ _ ☐ _ _ _

What does Alcatraz mean in Spanish? _____

Alcatraz has a nickname. Unscramble the letters below. Write its nickname on the lines following.

H　E　T　　C　O　R　K 　　　_ _ _　_ _ _ _
2　3　1　　6　5　4　7　　1　2　3　　4　5　6　7

Ellis Island

Name _____

Ellis Island is in New York Harbor. It was the place most people coming to the United States for the first time had to go before being allowed to stay in America. They were examined there by doctors and checked to be sure they had no criminal record. If they were found to have a disease or to have something wrong with their past, they were sent back to the country from which they had come.

Ellis Island is no longer used as an immigration station, but it stands as a symbol of the time when hundreds of foreigners passed through immigration. Thirty or more buildings remain on Ellis Island near ruin. They are being restored and should be open for the public to visit sometime after 1990.

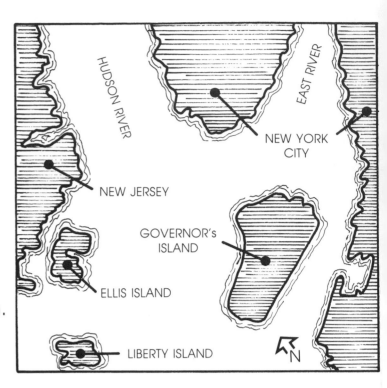

America is a melting pot. Originally Americans came from other countries besides the United States. The names of many countries from which the immigrants came are scrambled below. Unscramble them to learn the names of some of these countries.

A Y I L T   _____
3 5 1 4 2

P N A O D L   _____
1 5 4 2 6 3

M A N R O I A   _____
3 4 5 1 2 6 7

G R U N A H Y   _____
4 6 2 3 5 1 7

L I E A R D N   _____
4 1 3 5 2 7 6

N E F A R C   _____
4 6 1 3 2 5

S I R S U A   _____
3 5 1 4 2 6

Y N G R E M A   _____
7 6 1 3 2 4 5

G L E D N A N   _____
3 4 1 7 6 5 2

O C L H Z S K O C V A A E I   _____
9 4 8 5 2 7 12 6 1 10 14 11 3 13

Appomattox Court House

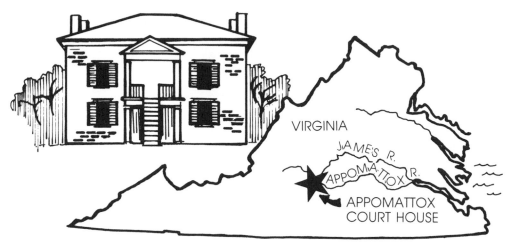

VIRGINIA
JAMES R.
APPOMATTOX R.
APPOMATTOX COURT HOUSE

Appomattox Court House was a small village of about 150 people in Virginia. It had a few homes, stores, lawyers' offices, a tavern and a courthouse. It was the county seat for Appomattox County, a farming area. On April 9, 1865, General Robert E. Lee, commander of the North Virginia Army, met with General Ulysses S. Grant, commander-in-chief of the Union army, in Wilmer McLean's farmhouse. Lee surrendered his men to Grant. Grant only asked that the Confederate men not take up arms again against the United States. He freed them to return to their homes. When news of Lee's surrender reached the Confederate armies still fighting out in the fields, the Confederate soldiers put down their weapons there and went back to their homes, too. The Civil War was over. Today Appomattox Court House is a national historical park open to the public.

Write either **TRUE** or **FALSE** in front of each statement about Appomattox Court House.

_____ Appomattox Court House was a small village.
_____ It is located in Virginia.
_____ Grant surrendered to Lee.
_____ Lee was commander of the North Virginia Army.
_____ The Union won the Civil War.
_____ Lee and Grant met in an old estate.
_____ The courthouse was the county seat.
_____ Grant was the commander-in-chief of the Union armies.
_____ Lee surrendered to Grant in August.
_____ After the Civil War, the Confederate soldiers were put in jail.
_____ Today Appomattox Court House is open to the public.
_____ Most of the people in Appomattox County were farmers.
_____ Wilmer McLean had a farmhouse.
_____ When Lee surrendered, all fighting stopped immediately.
_____ There were a few homes and offices in Appomattox Court House.

Independence Hall

Independence Hall was not always called by that name. It was built in Philadelphia as the state house for Pennsylvania and was called the Old State House. Its purpose was for Pennsylvania's lawmakers to meet there.

In 1775 America's early statesmen used the Old State House when they met to write the Declaration of Independence. Eleven years later the Constitution of the United States was written there. The Liberty Bell once hung in its tower. In 1818 Philadelphia bought the Old State House. In 1824 General Lafayette renamed it Independence Hall.

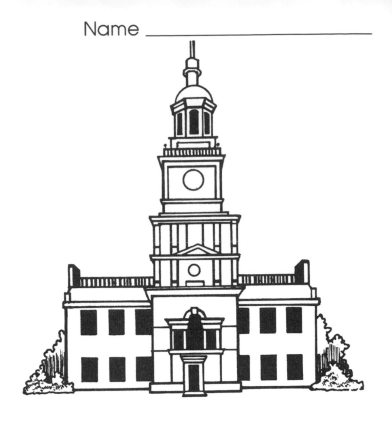

Today it is a National Historical Park. Visitors from all over the world come to see the "Shrine of Independence" where the United States government got started.

Circle every third letter below to learn one thing visitors will see when they visit Independence Hall. Write it out on the lines at the bottom of the page.

EGTIEHMQE   INSXUIBW LAOVZDEMCR   PUILKNJHKGFSDATQWAERNTYD

MZUXCSVBENMD   GMBPTY   WRTPUHIEE   QRSPUIVXGAINVRESDRABS

PIODAF   VIBPPOSRTVGH   VYTWIHQPE

IEDGHEPACXYLMBAEDRZYAOGTUNIVFOMSN   PWOVHF

OPINMNPADIEEBYPWSEXBNPPDAIECWNIQCQUE   ORAXFNOID   FGTBUHMAE

BACKEOLDNRESUPTRAINOTRQUYATPAIMBOANN

_____

_____

_____

_____

Presidential Homes

Read about two presidential homes and connect the dots to see how each one looks.

Mount Vernon was not always called Mount Vernon. George Washington's father built the main house in the 1730's and called the plantation "Little Huntington Creek". Mount Vernon was built in stages from 1742 to 1787. Washington's half brother, Lawrence, named it Mount Vernon when he headed the estate. When Lawrence died, Mount Vernon became Washington's. By the time Washington became President, the estate had five working farms and several outbuildings.

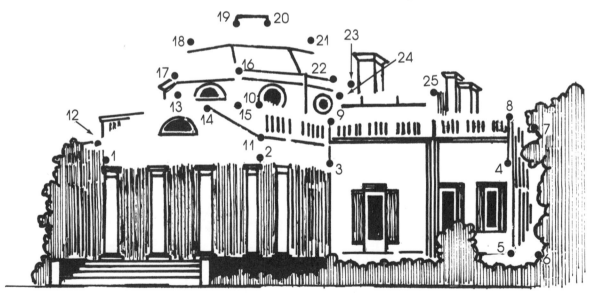

When Thomas Jefferson was fourteen years old, his father died and left him 2,750 acres in Virginia. Nine years later, in 1768, Jefferson designed and began to build Monticello on a hilltop. In Italian Monticello means "little mountain". The house was not completed until 1809. It was filled with many of Jefferson's inventions, such as a dumb waiter, calendar clock and a revolving desk. As in many homes like Monticello and Mount Vernon, the kitchens were not in the main structure. Do you know why?

The Statue of Liberty

The Statue of Liberty is a huge national monument that stands on Liberty Island in New York Harbor. It was given to the United States by France on July 4, 1884, as a symbol of friendship and liberty. France paid for the statue. The American people paid for the pedestal on which it stands.

The statue was designed and made in France. It was shipped to the United States in 214 boxes. The statue was put together and placed on its pedestal which had been built over old Fort Wood. The statue was completed in 1886.

The statue became a national monument in 1924. When the statue was 100 years old, there was a big birthday party on July 4, 1986.

Color the statue green.
Color the pedestal tan.

Unscramble the letters after each question below to find the answer. Write it on the line following the scrambled word.

What was the name of the fort upon which the Statue of Liberty stands?

ORFT ODOW _____

Who gave the statue to the United States? **CNREFA** _____

On what island does the statue stand? **BIRLYET** _____

In what Harbor is Liberty Island? **WNE ROYK** _____

What country paid for the pedestal? **RMAEACI** _____

The Liberty Bell

The first Liberty Bell was ordered from London in 1752 by the Pennsylvania legislature in observance of their fifty years of liberty as a colony. The first time the bell was rung, it cracked. A new bell was recast by a Philadelphia foundry using the same metal. The second bell did not have a good tone, so a third bell was made.

The bell was the official town bell. It rang to call town meetings, announce the end of wars and deaths of great men. It also served as a fire alarm. It rang to announce the signing of the Declaration of Independence in 1776.

During the Revolutionary War, the Liberty Bell was moved to Allentown, Pennsylvania, so the enemy could not be able to seize it and melt it down for ammunition. In 1778 when the British had gone from Philadelphia, the bell was returned to its place in the State House and rang again. It rang every Fourth of July until 1835. Not until the 1840's was it called the Liberty Bell.

Work the math problems below to find what the other names were for the bell. Then look to see what letter goes with each answer. Write the letter below each answer.

A=1 B=2 C=3 D=4 E=5 F=6 G=7 H=8 I=9
J=10 K=11 L=12 M=13 N=14 O=15 P=16 Q=17
R=18 S=19 T=20 U=21 V=22 W=23 X=24 Y=25 Z=26

$$\begin{array}{cccccccccccccc} 8 & 16 & 10 & 4 & 9 & 2 & 20 & 8 & 13 & 15 & 7 & 2 & 6 & 9 & 12 \\ +7 & -4 & -6 & +5 & +5 & +2 & -15 & +8 & -8 & -1 & -3 & +3 & +8 & -6 & -7 \end{array}$$

___ ___ ___   ___ ___ ___ ___ ___ ___ ___ ___ ___ ___ ___ ___,

$$\begin{array}{cccccccccccccccccc} 19 & 4 & 9 & 10 & 11 & 6 & 18 & 11 & 3 & 13 & 14 & 20 & 1 & 8 & 14 & 4 & 16 \\ -4 & +8 & -5 & +9 & +9 & -5 & +2 & -6 & +5 & +2 & +7 & -1 & +4 & -6 & -9 & +8 & -4 \end{array}$$

___ ___ ___   ___ ___ ___ ___ ___   ___ ___ ___ ___ ___   ___ ___ ___ ___   OR

$$\begin{array}{ccccccccc} 11 & 0 & 15 & 6 & 9 & 10 & 10 & 12 & 9 \\ -9 & +5 & -3 & +6 & +6 & -4 & +10 & -4 & -4 \end{array}$$

___ ___ ___ ___   ___ ___   ___ ___ ___

$$\begin{array}{cccccccccc} 9 & 11 & 11 & 20 & 7 & 10 & 15 & 12 & 17 & 7 \\ +9 & -6 & +11 & -5 & +5 & +11 & +5 & -3 & -2 & +7 \end{array}$$

___ ___ ___ ___ ___ ___ ___ ___ ___ ___

The Star-Spangled Banner

Name _____

The Star-Spangled Banner was written by Francis Scott <u>Key</u>. Key had permission from President <u>Madison</u> to meet with the British to try and have a friend that had been captured freed. Key sailed into Baltimore's harbor to meet the <u>British Admiral</u> who agreed to set Key and his friend free after the attack on Baltimore.

Key watched the flag above Fort <u>McHenry</u> all through the <u>day</u>, September 13, 1814. That night he listened to bombs bursting in the air. At <u>dawn</u> on September 14th, when he saw the flag still flying over Fort McHenry, he knew the British had lost and that Baltimore was still a city. He was so happy that he <u>wrote</u> a <u>poem</u> to express how he felt. The poem was set to <u>music</u> written by an <u>Englishman</u> for a different poem.

Fill in the missing words below. They are underlined above. The circled letters in your answers will spell out the first five words of the Star-Spangled Banner.

__ __ __ __ __ ◯ __ was the President when the Star-Spangled Banner was written.

Francis Scott Key went to meet

the __ __ __ __ ◯ __ ◯ __ __ __ __ __ __.

The battle began during the __ __ ◯.

The flag above Fort __ ◯ __ __ __ __ __ lasted through the battle.

At __ ◯ __ ◯ it was still there.

__ __ ◯ was so happy that he __ __ ◯ __ __ a poem.

It was set to __ ◯ ◯ __ __ written by

an ◯ __ __ __ __ __ __ __ __ for a different __ __ ◯ __.

Write the first five words of the Star-Spangled Banner.

_____

Uncle Sam

Name _____

Uncle Sam is not a real person. He is a symbol of the United States. He often appears on posters asking the American people to do something good for their country. Although he has been around for over one hundred fifty years, Congress did not recognize him as a national symbol until 1961.

First draw Uncle Sam below by connecting the dots. Then color him by using the color code.

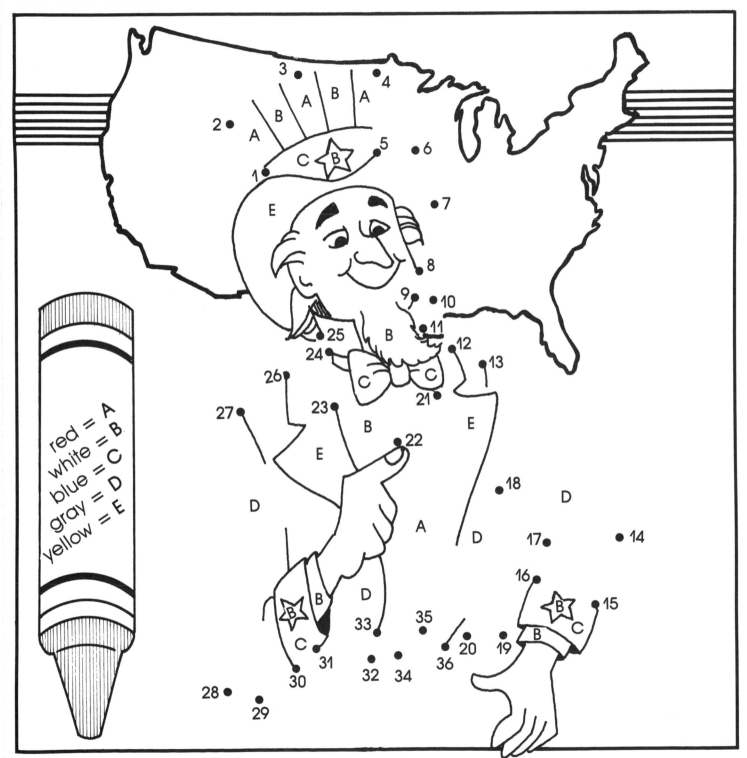

89

© 1992 Instructional Fair, Inc.

The Flag of the United States

Name _____

The Continental Congress left no record of why they chose red, white and blue as the colors of the flag. The stripes in the flag were to stand for the original colonies. In 1777 Congress said the flag should have thirteen stars, but it did not say how they should be arranged. Every time a state was added to the Union, another star was added to the flag. The stars had no special arrangement until 1912. Since then the President has ordered how the stars are to be arranged.

Connect the dots in the boxes below. Color each flag.

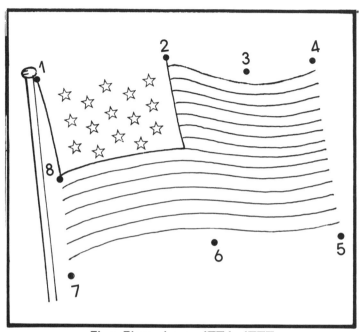

The Flag from 1776–1777

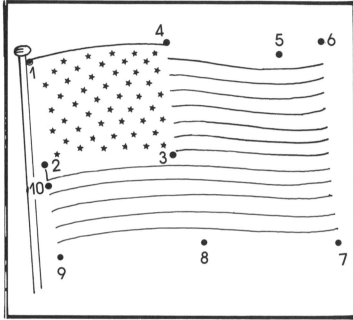

The Flag Now

How are they alike? _____

How are they different? _____

Over the years the United States flag has had many names. Unscramble the letters below and write them in order on the lines following to learn some of the flag's other names.

L	O	D
2	1	3

L	Y	O	R	G
5	8	6	7	4

___ ___ ___   ___ ___ ___ ___ ___

A	R	S	T	S
3	4	1	2	5

A	D	N
6	8	7

T	I	E	R	P	S	S
10	12	14	11	13	9	15

___ ___ ___ ___ ___   ___ ___ ___   ___ ___ ___ ___ ___ ___ ___

T	A	R	S	-	N	G	S	P	E	A	L	D
2	3	4	1		8	9	5	6	11	7	10	12

N	E	R	B	N	A
15	17	18	13	16	14

___ ___ ___ ___ ___ ___ ___ ___ ___ ___ ___ ___   ___ ___ ___ ___ ___ ___

The Pledge of Allegiance

Name_____

The Pledge of Allegiance is a promise to be true to the United States. It was first recited in 1892 by children saluting the flag. President Benjamin Harrison had asked there be patriotic celebrations in America's schools to mark the 400th anniversary of the country's discovery. Francis Bellamy wrote the original pledge. It was changed two or three times before it had the wording we use today.

Cross out every third letter below beginning with the third one. The letters that remain will spell out the Pledge of Allegiance. The lines (—) between the letters are to help divide the words. Write the pledge on the lines at the bottom of the page.

I—PALEXDGME—AFLLUEGRIARNCNE—THO—TIHEN—FLSAGT—OFE—THAE—

UHNIBTERD—SCTAPTEOS—OJF—AKMEGRIDCAS—ANOD—TRO—TSHEW—

RESPUZBLAICY—FOER—WEHIRCHS—ITS—STRANLDSI—ONCE—NEATRIOUN—

UENDLERS—GOLD—IENDRIVAISSIBLLES—WIATHY—LIWBEVRTEY—ALNDE—

JUXSTRICEE —FLORT—ALIL

_____

_____

_____

_____

_____

_____

The Declaration of Independence

Name _____

The Declaration of Independence is a **document** written in 1776 in which America's thirteen **colonies** declared their freedom from British **rule**. England had been deciding many of the laws for the new country and had been taxing the colonists too much.

The colonists were very unhappy. The Declaration of Independence lists their complaints against the King of England and gives good reason for the breaking away of the colonies from England.

The **Second** Continental Congress appointed a committee to write the declaration. **Thomas Jefferson** was the main author of the document. Others on the committee were **Robert Livingston, John Adams, Roger Sherman** and **Benjamin Franklin**. The Declaration of Independence is signed by **fifty-six** delegates. John Hancock signed it **First** on July 4, 1776, because he was President of the Second Continental Congress.

Fill in the blanks below with the underlined words from above. Write out the circled letters in order in the box at the bottom of the page. They will spell the name of the holiday on which we celebrate the birth of our country.

1. __ __ __ __ __ __ __ __ Ⓞ __ __ __ __ __ __ __ was the main author of the Declaration of Independence.

2. The __ __ __ Ⓞ __ __ Continental Congress appointed a committee to write the Declaration of Independence.

3. The colonists broke away from British __ Ⓞ __ __.

4. Ⓞ __ __ __ Ⓞ __ __ __ __ __ __ __ __ __ __ __ __, __ __ __ __ __ __ Ⓞ __ __ __ __ and __ Ⓞ __ __ __ __ __ __ __ were other authors of the Declaration of Independence.

5. The Ⓞ __ __ __ __ to sign the Declaration of Independence was John Hancock.

6. The fifth author of the Declaration of Independence was __ __ Ⓞ __ __ __ __ __ __ __ __ __ __.

7. The Declaration of Independence is a __ __ __ Ⓞ __ __ __ __ of freedom.

8. States were called __ __ Ⓞ __ __ __ __ __ in 1776.

9. __ __ __ __ Ⓞ-__ __ __ delegates signed the Declaration of Independence.

The Constitution of the United States

Name _____

The Constitutional Convention met in Philadelphia in 1787. Fifty-five delegates from twelve of the thirteen states were there. George Washington was President of the Constitutional Convention. The Constitution was written between May 25 and September 17, 1787. Only thirty-nine of the delegates signed it. William Jackson, secretary to the convention, signed as witness to the signatures. After it was signed, copies were made and sent to all the states for their approval. The Constitution was ratified first by Delaware on December 7, 1787.

The United States became a united nation with the signing of the Constitution. The Constitution's seven articles and twenty-six amendments are the basis of our laws. The Constitution lists the three branches of government—executive, legislative and judicial. It lists the rights of the people. Today the Constitution is on display, under glass, in Washington, D.C. in the National Archives Building.

Answer **TRUE (T)** or **FALSE (F)** to the following statements.

____ All states attended the Constitutional Convention.

____ George Washington was President of the U.S. in 1787.

____ The Constitutional Convention met in Philadelphia.

____ Forty delegates signed the Constitution.

____ The Constitution established the country's laws.

____ The original copy of the Constitution is in the National Archives Building.

____ All the delegates signed the Constitution.

____ The states had to approve the Constitution.

____ The Constitution was written in three days.

____ On June 21, 1788, the Constitution was approved.

____ The Constitution has seven articles and twenty-six amendments.

____ William Jackson was the convention's secretary.

____ The formation of the executive, legislative and judicial branches are outlined in the Constitution.

____ The original Constitution is in Philadelphia now.

____ Several copies of the Constitution were made by the Constitutional Convention.

Bill of Rights

Congress OF THE United States

James Madison presented twelve amendments to the first Congress in 1789. Congress passed ten of them. They are called the Bill of Rights. They became law December 15, 1791. In 1940, President Roosevelt declared December 15th Bill of Rights day.

The Bill of Rights describes freedoms every American has. Some of the rights Americans have are freedom of speech, press, religion and assembly. Every American may say and write what he or she wants, and pray and meet where he or she wants. The Bill of Rights guarantees freedom for all Americans—something that we perhaps just take for granted.

Use the underlined words above to fill in the crossword puzzle.

Across
2. Last name of man responsible for Bill of Rights
6. One of the rights
8. One of the rights
9. One of the rights

Down
1. Number of amendments that are part of the Bill of Rights
3. Month they became law
4. What every American is guaranteed with the Bill of Rights
5. First name of man responsible for the Bill of Rights
7. One of the rights

The Great Seal of the United States

Name _____

The United States government adopted the Seal of the United States on June 20, 1782. It used the seal on important documents. The American eagle, with a shield on its chest, represents self-reliance. The stripes on the shield come from the flag of 1777. The blue at the top of the shield symbolizes all branches of government. The eagle holds thirteen olive branches in its right talon and thirteen arrows in its left showing its desire to live in peace, but with the capability of waging war. The yellow banner in the eagle's beak has Latin words on it meaning "One (nation) out of many (states)." Above the eagle's head is a circle with thirteen stars inside a golden circle breaking through a cloud.

Color the Great Seal of the United States using the color key to the left.

1 – blue   5 – yellow
2 – brown   6 – white
3 – green   7 – grey
4 – red   8 – black

This is the face of the seal. There is a reverse side. It has never been used as a seal, but it can be found on the back of a dollar bill. Circle the reverse side of the seal on the left.

The White House

The official residence of the President of the United States has not always been called the White House. It was first called the President's House, then the Executive Mansion. Finally in 1901, Theodore Roosevelt gave it its name, the White House.

One and a half million visitors are given guided tours of the White House's reception rooms every year. They are not allowed above the first floor. The President and his family live on the second floor, and the third floor has guest rooms and servants' quarters. Offices have been added over the years to the East and West wings of the White House.

The answer to each question below is spelled out in the circle next to the question. Begin with the arrow. Go around to the right to find the answer to each question. Write it on the line next to the circle.

1. Who was the first President to live in the White House?

_____

2. Which President had to leave the White House when the British set fire to it?

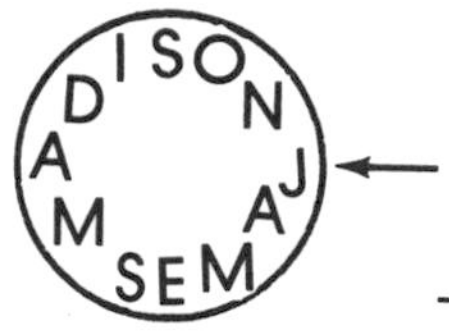

_____

3. In what city is the White House?

_____

4. What is the name of one of the rooms visitors may see?

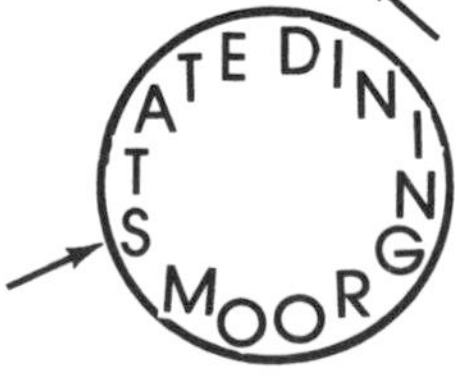

_____

5. On what street is the White House?

_____

Who lives in the White House now? _____

The United States Capitol

The United States Capitol is a symbol of the nation and a government office building where Congress meets to make the nation's laws. The building is made of marble. The dome is made of iron and painted white to look like the rest of the Capitol. The Capitol was built in 1800, burned in 1814 and rebuilt. In 1863 the Capitol received a new dome. A statue of a woman named Freedom was placed on top. The distance from the top of the statue to the ground is about 300 feet.

Connect the dots to make a picture of the Capitol.

— north ⟶

The Senate meets in the north wing of the Capitol. Circle it. The House of Representatives chamber is in the south wing. Draw a line under it.

The center part of the Capitol is called the Great Rotunda. Mark it with an X.

The Supreme Court

Name_____

When the Constitution was written, it provided for three branches of government—the Executive, Legislative and Judicial. The Supreme Court heads the Judicial branch. It is the highest court in the land. It is responsible for carrying out the laws as written in the Constitution.

Congress determines the number of judges, or justices, that sit on the Supreme Court. Since 1869 there have always been nine. The justices are appointed by the President and approved by the Senate. In this way no one branch has all the power. Once a justice is appointed, he or she may hold the position for life. The Supreme Court meets from October to June every year. For 135 years it met in the Capitol. Since 1935 it meets in its own building.

There is an inscription over the entrance to the Supreme Court building which states the Court's belief and duty. To learn what it says, find the answer to each problem. Look to the left to see what letter goes with each answer. Write the letter under the number that represents it.

A = 4	N = 15
B = 3	O = 14
C = 2	P = 13
D = 1	Q = 20
E = 8	R = 19
F = 7	S = 18
G = 6	T = 17
H = 5	U = 26
I = 12	V = 25
J = 11	W = 24
K = 10	X = 22
L = 9	Y = 23
M = 16	Z = 21

$$\begin{array}{cccccccccccc}
3 & 14 & 13 & 9 & 17 & & 7 & 30 & 23 & 9 & 4 & 9 & 14 \\
+5 & +6 & +13 & -5 & -8 & & +4 & -4 & -5 & +8 & +8 & -7 & -6 \\
\hline
\rule{0pt}{1em} & & & & & & & & & & & & \\
\end{array}$$

$$\begin{array}{ccccccccc}
17 & 7 & 1 & 19 & 25 & & 5 & 13 & 12 \\
+9 & +8 & +0 & -11 & -6 & & +4 & -9 & +12 \\
\hline
\rule{0pt}{1em} & & & & & & & & \\
\end{array}$$

The inscription over the entrance to the Supreme Court building reads _____ .

The Gettysburg Address

Name _____

On November 19, 1863, four months after the bloodiest battle of the Civil War, Lincoln delivered his famous Gettysburg Address on the site of the battle. The purpose of his speech was to dedicate a part of the battlefield as a cemetery. Then the over 7,000 soliders from both the North and South who had lost their lives could be buried. He worked hard to make the speech perfect. He chose his words carefully because the Civil War was still being fought. He wrote five different versions of it. Only the last one is signed. A copy of it is carved in stone inside the Lincoln Memorial.

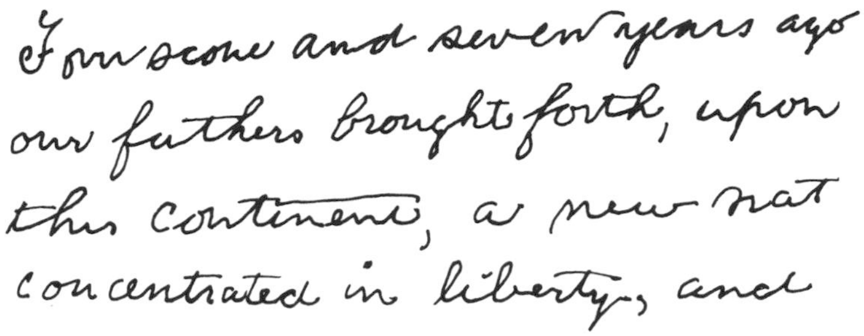

Four score and seven years ago our fathers brought forth, upon this continent, a new nat concentrated in liberty, and

Cross out every other letter below beginning with the second one. The letters that remain will spell out how many words were in the Gettysburg Address.

TWWEOX   HRUNNODTRGEIDT   STERVOENNUTVYE — THWIOL

There were _____ words in the Gettysburg
Address.

Do it again to learn how many minutes it lasted.

TPWIOG   MUIGNWUTTHEDSC

The speech lasted _____.

Do it one more time. The letters that remain will spell the first six words of the Gettysburg Address.

FPOEUVRX   SPCQOTRAEN   AVNQDE   SWECVLEBNY   YAEAAFRGSM   ARGUOL

The first six words of the Gettysburg Address are:

_____.

One score = 20. Four score = 80. 80 + 7 = _____ How many years was he talking about? _____

The Vietnam Veterans' Memorial

Name _____

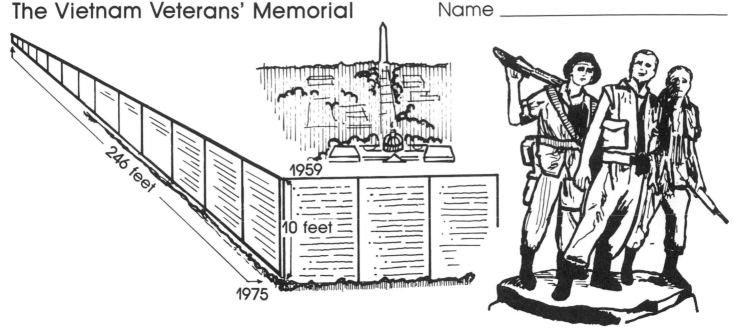

246 feet

1959

10 feet

1975

The Vietnam Veterans' Memorial honors all men and women of the United States armed forces who served in the Vietnam War from 1959 to 1975. Congress gave the land for it, but the memorial was built with contributions from business and civic groups and more than 275,000 private citizens. It cost $7,000,000. The memorial was dedicated in 1982. A sixty-foot flagpole and a three-figured sculpture by Frederick Hart were added to the memorial site in 1984.

The memorial is on the Mall in Washington, D.C. set between the Washington Monument and the Lincoln Memorial. The names of the 58,022 men and women who lost their lives or remain missing are carved in its two black granite walls.

Fill in the cross word with the correct numbers.

ACROSS
2. Year Vietnam War began
4. Year sculpture and flagpole were added
5. Number of citizens who helped build memorial with dollars
7. Height of flagpole in feet
9. Number of names carved in walls

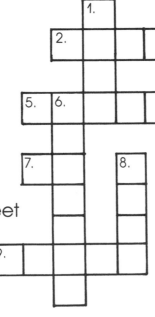

DOWN
1. Year Vietnam War ended
3. Length of each wall in feet
4. Height of memorial where two walls join in feet
6. The millions of dollars it cost
8. Year memorial was dedicated

Arlington National Cemetery

Name _____

Arlington National Cemetery is in Virginia just across the Potomac River from Washington, D.C. There are over 200,000 Americans buried there. It is run by the United States Army. John F. Kennedy, the 35th President of the United States, Joe Louis, a world heavyweight champion, and Adm. Robert Peary, discoverer of the North Pole, are some of the better known to be buried there along with thousands of military personnel.

Besides the thousands of grave markers, there are several special monuments in the cemetery. To learn what some of them are, find the answer to each math problem. Look to the left to see what letter goes with each answer. Write the letter under the number that represents it. Under that will be a fact about the monument.

A = 2
B = 1
C = 4
D = 3
E = 6
F = 5
G = 8
H = 7
I = 10
J = 9
K = 12
L = 11
M = 14
N = 13
O = 16
P = 15
Q = 18
R = 17
S = 20
T = 19
U = 22
V = 21
W = 24

```
  2     8     7    10     9     1     8     8     1    20     3
+ 2   + 8   + 6   - 5   - 3   + 2   - 2   + 9   + 1   - 1   + 3
____  ____  ____  ____  ____  ____  ____  ____  ____  ____  ____
```
____ ____ ____ ____ ____ ____ ____ ____ ____ ____ ____ MEMORIAL

Five hundred soldiers from the Civil War are buried around this memorial.

```
 18     6     6     8    12
- 4   - 4   + 4   + 5   - 6
____  ____  ____  ____  ____
```
____ ____ ____ ____ ____ MEMORIAL

It is the mast of the U.S.S. Maine Battleship that was blown up to start the Spanish American War in 1898.

```
  4    13    10     4     9    18     9     0    10     8    10
+ 9   - 7   + 9   + 3   - 3   - 1   + 2   + 2   + 3   - 5   +10
____  ____  ____  ____  ____  ____  ____  ____  ____  ____  ____
```
____ ____ ____ ____ ____ ____ ____ ____ ____ ____ ____

```
  8     7    12    13     8     3     9    11
- 4   - 5   + 5   - 3   + 3   + 8   + 7   + 2
____  ____  ____  ____  ____  ____  ____  ____
```
____ ____ ____ ____ ____ ____ ____ ____

These 49 bells were a gift from the people of the Netherlands.

The Tomb of the Unknowns

The Tomb of the Unknowns in Arlington National Cemetery was originally called The Tomb of the Unknown Soldier. An unknown American soldier from World War I was laid to rest there on Armistice Day, November 11, 1921. In 1958, the name of the memorial was changed to The Tomb of the Unknowns when an unidentified serviceman from World War II and the Korean War were buried there. On Memorial Day, May 28, 1984, a fourth unknown soldier from the Vietnam War was laid to rest between the World War II and Korean heroes.

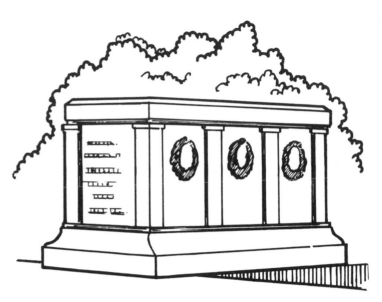

The Tomb is guarded 24 hours a day, 365 days a year. The guards belong to the Honor Guard Company, Third U.S. Infantry. The guards have a ritual based on the number 21. It is symbolic of a 21 gun salute—the highest salute given. The guard faces the tomb for 21 seconds, turns and pauses for 21 seconds, and then walks 21 steps, faces the tomb and repeats going in the opposite direction. Back and forth he goes until the guard changes—every half hour in the summer; every hour in the winter. Over 3,000,000 visitors enjoy this ceremony annually.

There are some words written on the tomb. Cross out every other letter below starting with the second one. Write the remaining letters on the blanks to learn what is written on the tomb.

HSEARBEL   REEASRTESC   IUNT   HEOLNIOTRTELDE   GBLEOARSYT

ASNB   AUMTETRHIECYACNA   SNOBLEDVIEERRY   KVNEORWYNM

BEUATN   TAON   GDOSDN

" __ __ __ __   __ __ __ __ __   __ __   __ __ __ __ __ __

__ __ __ __ __   __ __   __ __ __ __ __ __ __

__ __ __ __ __ __ __   __ __ __ __ __ __

__ __   __ __ __ __ "

Answer Key

Alabama
the Heart of Dixie

Name _____

Becoming the 22nd state in 1819, Alabama has experienced great change. Dominated agriculturally by cotton for so long, its farm income today consists primarily of such things as broiler chickens, beef cattle, soybeans, corn, peanuts and strawberries.

And, dominated by the steel industry for so long, its manufacturing income now includes textiles, aluminum, paper and chemical products.

Race relations have long been a problem, but blacks are beginning to play a greater role in politics. This change was begun by the famous Martin Luther King, Jr.

yellowhammer

MORE TO LEARN

Word search grid:

```
B C A T T L E C A
R H S E W H E P P
O I T X V C D A E
I C O T T O N P A
L K C I N R L E N
E E O L X N L R U
R N R E W H E A T
S S O Y B E A N S
C H E M I C A L M
```

Circle the various crops, products, etc. mentioned above in the WORD SEARCH. There are ten, going → or ↓. Anytime a word is circled that has a box around a letter, write that letter below.

A E C A L L W

Now, unscramble the letters to discover the last name of the four-time governor who tried to stop school integration.

W A L L A C E

CAN YOU FIND . . . What baseball player from Alabama holds the home run record? Hank Aaron

Page 1

Alaska
the Last Frontier

Name _____

Alaska, the 49th state, ranks first in size and last in population in the nation! Juneau, its capital, is the nation's largest city in area — over 3,000 square miles — but has only 19,000 people.

Alaska touches no other state. Instead, it is bordered on the north by the Arctic Ocean, on the south by the Pacific Ocean, on the east by Canada and on the west by the Bering Sea. At one point it is only 5½ miles across the Bering Strait to Russia.

The vast untold natural resources and incredible beauty of this "Land of the Midnight Sun" make it truly America's last frontier.

Label the bodies of water, bordering country, nearby country and state capital on the map.

MORE TO LEARN

Mark each statement TRUE OR FALSE. On another sheet of paper, rewrite each FALSE statement so it is TRUE.

willow ptarmigan

- F Alaska touches one other state.
- T The state ranks first in size in the nation.
- F Anchorage is the largest city and the state capital.
- F Juneau has the largest population of any city in the nation.
- T Alaska ranks last in population of the fifty states.
- F Mt. McKinley is the highest peak in the world.
- T Canada borders Alaska on the east.
- T The state is bordered by two oceans.

CAN YOU FIND . . . What was the 1867 purchase of Alaska for $7,200,000 called by Americans? Seward's Folly

Page 2

Arizona
the Grand Canyon State

Name _____

cactus wren

Arizona, the 48th state, has a history greatly influenced by its need for water. The first irrigation of its desert was by Indians nearly 800 years ago. Without irrigation, half the state would be a desert.

The warm, dry climate has made Arizona a rapidly-growing state. Most of the people live in or near Tucson and Phoenix, the capital and largest city.

A history of Mexican and Indian influence is quite apparent in Arizona's culture.

saguaro blossom

On the map, label the following: state capital, second largest city, bordering country and river running through the Grand Canyon.

MORE TO LEARN

Write the number of the phrase in Column B that describes who or what each is in Column A. Think logically!

A		B
Sandra Day O'Connor	5	1. Indian chief who kept fighting
Tombstone	4	2. Located at Lake Havasu City (but from England)
Geronimo	1	3. River running through the Grand Canyon
Barry Goldwater	8	4. Where Wyatt Earp won fame
Colorado	3	5. First female U.S. Supreme Court Justice
Navajo	7	6. Largest cactus in U.S.
Saguaro	6	7. Largest Indian tribe in Arizona
The London Bridge	2	8. Father of Arizona Republican Party

CAN YOU FIND . . . What are the names of two national monuments in Arizona? Organ Pipe Cactus, Pipe Spring, Sunset Crater, etc.

Page 3

Arkansas
the Land of Opportunity

Name _____

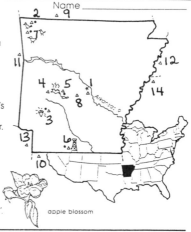

Arkansas, the 25th state, is a land rich in natural resources, both above and beneath the soil. Its lowlands have rich soil for farming. Its plateau area is suitable for raising hogs, cattle and broiler chickens (the country's leader). One quarter of the state is covered with rich timber.

Beneath the soil is found nearly all the nation's bauxite, the only diamond mine, and deposits of petroleum and natural gas.

The state's natural beauty attracts millions of tourists each year to this "Land of Opportunity".

apple blossom

MORE TO LEARN

Use the pictures on the map and the clues below to identify the towns, cities, bordering states and points of interest. Write the numbers next to the △'s on the map.

1. Little Rock, the state capital
2. Pea Ridge, site of a Civil War battle
3. Murfreesboro, town near only diamond mine in the U.S.
4. Hot Springs National Park
5. Town of Hot Springs
6. El Dorado, site of first oil discovery in the state
7. Springdale, in center of broiler chicken area
8. Bauxite, site of first aluminum ore mine
9. Missouri - north
10. Louisiana - south
11. Oklahoma - west
12. Tennessee - northeast
13. Texas - southwest
14. Mississippi - southeast

mockingbird

CAN YOU FIND . . . Who was the Arkansas woman that was the first female elected to the U.S. Senate? Hattie Caraway

Page 4

Answer Key

California
the Golden State

California, the 31st state, is almost beyond comprehension! In America, this state has the largest population, the most goods produced, the highest agriculture output, the tallest and oldest living things, the largest city (Los Angeles) — the list is endless.

Also a land of great natural beauty, California's Yosemite National Park contains Ribbon Falls, the highest waterfall in North America, and Mount Whitney, the highest point in the U.S. south of Alaska. Yet its Death Valley is the lowest point in North America! Its 840 miles of coastline varies from steep cliffs to sandy beaches with two great natural harbors — San Francisco and San Diego.

Label the cities and points of interest on the map.

MORE TO LEARN

Locate the following in the WORD SEARCH:

- The bordering states
- The two natural harbors
- Lowest point in North America
- Highest waterfall in North America
- State capital
- Highest mountain in U.S. south of Alaska
- A national park

Word search grid containing: SANTS NEVADA, YOSEMITE, WHITNEY, DEATH, VALLEY, OREGON, SACRAMENTO, FRANCISCO

CAN YOU FIND... Who was the famous English sea captain who sailed around the world claiming what is now California for England in 1579? **Sir Francis Drake**

Page 5

Colorado
the Centennial State

Colorado, the 38th state, is the highest state in the nation. It has the highest road in the U.S. and the world's highest tunnel for vehicles.

The Rocky Mountains, cause of the transportation problems, also attract tourism, provide vast mineral resources and are the source of water for the plains area.

A fast-growing state, Colorado's need for water storage has been a severe and ongoing problem. Water is the lifeblood of the people.

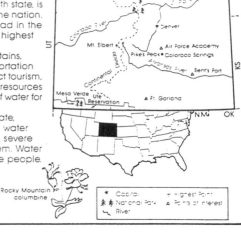

lark bunting

Rocky Mountain columbine

MORE TO LEARN

Which rivers flow east? **Arkansas, S. Platte** west? **Colorado**
Which state borders on the east? **KS** west? **UT** northeast? **NE**
southeast? **OK** northwest? **WY** southwest? **NM**
Colorado's mile-high capital is **Denver**
The oldest residents of Colorado are the **Ute** Indians who live on a reservation in the southwest corner of the state.
The highest mountain is **Mt. Elbert**. The second highest is **Pike's Peak**
The U.S. Air Force Academy is near **Colorado Springs**
Kit Carson long ago commanded the historic posts **Bent's Fort** and **Ft. Garland**
The two national parks are **Mesa Verde** and **Rocky Mountain**

CAN YOU FIND... Why is Colorado known as "the Centennial State"? **It became a state on the 100th anniversary of the Declaration of Independence.**

Page 6

Connecticut
the Constitution State

Connecticut, the 5th state, has a history of innovation. This trait accounts for its nickname because the Fundamental Orders, drafted in 1638, was the first constitution in the New World.

Industrialization got an early start with mass production and steel manufacturing beginning there. The first insurance for accidents and automobiles was written in Hartford, now known as the "Insurance City". And, Connecticut established the first American public school!

MORE TO LEARN

Using the WORD BANK, fill in the blanks to learn more about Connecticut events and people. The circled letters, when unscrambled, spell the last name of the man who first vulcanized rubber, Charles **Goodyear**

WORD BANK			
Oliver Cromwell	Nautilus	Litchfield	Mark Twain
Katherine Hepburn	Nathan Hale	Eli Whitney	hamburger

Nathan Hale, hung as spy during the Revolutionary War
Oliver Cromwell, the first American warship
Nautilus, first atomic submarine
Katherine Hepburn, famous actress
Mark Twain wrote Huckleberry Finn there.
Litchfield, site of America's first law school
Eli Whitney made the state the birthplace of mass production.
Hamburger, a favorite American food, first eaten there

CAN YOU FIND... What famous circus owner is from Connecticut? **P. T. Barnum**

Page 7

Delaware
the First State

The only state to belong to Sweden and Holland, England won the area that is known now as Delaware from the Dutch in 1664. About 100 years later, the American flag was first displayed in battle at the Revolutionary Battle of Cooch's Bridge. Named for Lord De La Warr, Delaware was the first state to ratify the new constitution in 1787.

Wilmington, the largest city, is the home of E. I. du Pont de Nemours and Co., the world's largest manufacturer of chemicals. The DuPont family has had great economic and political influence on the state for many years.

Write the names of the cities and points of interest labeled on the map.
Dover Wilmington Rehoboth Winterthur Museum Cooch's Bridge Battlefield

blue hen chicken

MORE TO LEARN

Dover, the state capital
E. I. duPont de **Nemours and Co.**, the world's largest chemical manufacturer
Lord De La **Warr**, for whom state was named
Atlantic Ocean, bordering on the east
Winterthur Museum, location of Art Conservation Project
American flag first displayed in battle at **Cooch's Bridge** September 3, 1777
Ratified the new constitution first!
England won the area from the Dutch in 1664.

CAN YOU FIND... What county in Delaware leads the nation in production of broiler chickens? **Sussex**

Page 8

Answer Key

Florida
the Sunshine State

Name _____

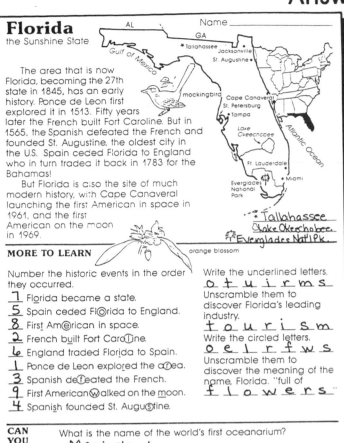

The area that is now Florida, becoming the 27th state in 1845, has an early history. Ponce de Leon first explored it in 1513. Fifty years later the French built Fort Caroline. But in 1565, the Spanish defeated the French and founded St. Augustine, the oldest city in the U.S. Spain ceded Florida to England who in turn traded it back in 1783 for the Bahamas!

But Florida is also the site of much modern history, with Cape Canaveral launching the first American in space in 1961, and the first American on the moon in 1969.

MORE TO LEARN

Number the historic events in the order they occurred.

- 7 Florida became a state.
- 5 Spain ceded Fl**o**rida to England.
- 8 First Am**e**rican in space.
- 2 French built Fort Car**o**line.
- 6 England traded Florida to Spain.
- 1 Ponce de Leon explored the a**r**ea.
- 3 Spanish de**f**eated the French.
- 9 First American **w**alked on the **m**oon.
- 4 Spanish founded St. Augu**s**tine.

Write the underlined letters.

o t u i r m s

Unscramble them to discover Florida's leading industry.

t o u r i s m

Write the circled letters.

o e l r f w s

Unscramble them to discover the meaning of the name, Florida. "full of

f l o w e r s "

CAN YOU FIND ...	What is the name of the world's first oceanarium? Marineland

Page 9

Georgia the Empire State

Name _____

Georgia, the 4th state, has come a long way in handling its long term racial problems. Two great leaders contributed much to integration — the late Martin Luther King, Jr. and Andrew Young, Ambassador to the United Nations in 1977.

Being an agrarian state, Georgia is now dominated by service industries (wholesale and retail trade). Farming is still important as it is a leading producer of tobacco, peaches and peanuts.

The state's most famous peanut farmer is Jimmy Carter — former President of the United States.

MORE TO LEARN

Use the code to learn some interesting people, places and events.

1 - F	2 - M	3 - P	4 - B	5 - Z	6 - L	7 - A	8 - T	9 - V
10 - C	11 - N	12 - W	13 - S	14 - D	15 - E	16 - R	17 - G	18 - J
19 - H	20 - K	21 - Q	22 - I	23 - U	24 - Y	25 - X	26 - O

Huge bird sanctuary

O K E F E N O K E E   S W A M P
26 20 15 1 15 11 26 20 15 15   13 12 7 3

Route of Cherokee nation to Oklahoma

T R A I L   O F   T E A R S
8 16 7 22 6   26 1   8 15 7 16 13

U. S. President who died at Warm Springs

F R A N K L I N   D.   R O O S E V E L T
1 16 7 11 20 6 22 11   14   16 26 26 13 15 9 15 6 8

Union General who captured Atlanta

W I L L I A M   S H E R M A N
12 22 6 6 22 7 2   13 19 15 16 2 7 11

First female U. S. Senator

R E B E C C A   L.   F E L T O N
16 15 4 15 10 10 7   6   1 15 6 8 26 11

Georgia woman who wrote Gone With the Wind

M A R G A R E T   M I T C H E L L
2 7 16 17 7 16 15 8   2 22 8 10 19 15 6 6

CAN YOU FIND ...	Who invented the cotton gin near Savannah in 1793? Eli Whitney

Page 10

Hawaii the Aloha State

Name _____

Hawaii, the 50th state, is a group of 132 islands formed from volcanic mountains. Diamond Head is one of its most famous extinct volcanoes.

An important naval base, Pearl Harbor, was the first part of the U.S. territory to be attacked in World War II.

The state's economy is heavily dependent on the service industries (nearly 90%), with tourism being the most important. Hawaii's culture is unique, with its colorful dress, famous luaus, hula dance and use of many polynesian terms such as "malihini" (newcomer). Its alphabet consists of only twelve letters with two consonants never coming together in a word.

List the seven inhabited islands in order of size. Hawaii, Maui, Oahu, Kauai, Molokai, Lanai, Kahoolawe

MORE TO LEARN

Mark each statement (T) TRUE or (F) FALSE. On another sheet of paper, rewrite each FALSE statement so it is TRUE.

- F Hawaii is a group of islands formed from active volcanic mountains.
- T This is the only state that was an independent monarchy.
- F Hawaii was the only part of U.S. territory to be attacked during World War II.
- F Diamond Head is an active volcano.
- T Hawaii was the last state admitted to the Union.
- F The most famous craft is the making of pottery products.
- T Service industries account for nearly 90% of the state's gross.

CAN YOU FIND ...	What was the name given to the islands in 1778 by the English sea captain James Cook? Sandwich Islands

Page 11

Idaho the Gem State

Name _____

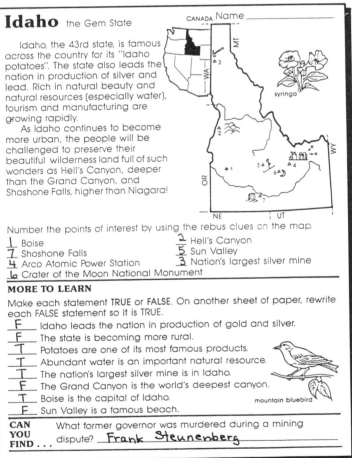

Idaho, the 43rd state, is famous across the country for its "Idaho potatoes". The state also leads the nation in production of silver and lead. Rich in natural beauty and natural resources (especially water), tourism and manufacturing are growing rapidly.

As Idaho continues to become more urban, the people will be challenged to preserve their beautiful wilderness land full of such wonders as Hell's Canyon, deeper than the Grand Canyon, and Shoshone Falls, higher than Niagara!

Number the points of interest by using the rebus clues on the map.

- 1 Boise
- 7 Shoshone Falls
- 4 Arco Atomic Power Station
- 6 Crater of the Moon National Monument
- 2 Hell's Canyon
- 5 Sun Valley
- 3 Nation's largest silver mine

MORE TO LEARN

Make each statement TRUE or FALSE. On another sheet of paper, rewrite each FALSE statement so it is TRUE.

- F Idaho leads the nation in production of gold and silver.
- F The state is becoming more rural.
- T Potatoes are one of its most famous products.
- T Abundant water is an important natural resource.
- T The nation's largest silver mine is in Idaho.
- F The Grand Canyon is the world's deepest canyon.
- T Boise is the capital of Idaho.
- F Sun Valley is a famous beach.

CAN YOU FIND ...	What former governor was murdered during a mining dispute? Frank Steunenberg

Page 12

Illinois
the Prairie State

Name _____

Illinois, the 21st state, is on one hand agrarian, leading the country in soybean and corn production. In contrast is Chicago, with the world's greatest railroad center, busiest airport and tallest building. There has always been a battle between the rural areas and Chicago for political control.

Illinois products are carried to many parts of the world, either down the Chicago River to Lake Michigan and the St. Lawrence Seaway, or the Illinois River which links with the Mississippi River and Gulf of Mexico.

Label Springfield (the state capital), the largest city, three rivers and the Great Lake.

MORE TO LEARN

Use the rhyming clues to fill in the blanks.

Th(e) world's busiest (a)irport, O'H **are** (rhymes with mare)
Chicago's gift to black l(e)adership, Jesse J**ackson** (rhymes with hack and ton)
Famou(s) railroad industrialist, George P**ullman** (rhymes with bull and tan)
(W)orld's largest mapmaker R **and** McN **ally** (rhymes with band and tally)
Famous U(n)ion general, Ulysses S. Gr**ant** (rhymes with plant)
Most i(m)portant natural resource, s**oil** (rhymes with toil)
Nation's (l)argest deposits of bituminous c**oal** (rhymes with foal)

Write the circled letters. **eaeswnml** Unscramble them to find the name of Lincoln's boyhood town which is now restored. **New Salem**

| CAN YOU FIND... | What President was born in Tampico, Illinois? **Ronald Reagan** |

Indiana
the Hoosier State

Name _____

Indiana, the 19th state, ranks 38th in the U.S. in area, but 12th in population. Its largest city and state capital is Indianapolis. The other large cities are Ft. Wayne in the northeast, Gary in the northwest and Evansville in the southwest. Heavily industrialized, manufactured products account for 84% of the total value of goods.

In South Bend is the famous Notre Dame University. In the south is huge Wyandotte Cave, and outside Indianapolis is the Indy 500 Raceway.

Label the cities and points of interest marked on the map.

MORE TO LEARN

Choose words from the WORD BANK to fill in the boxes.

NEW HARMONY	CORN	MEMORIAL
STUDEBAKER	SOIL	BEN-HUR
HARRISON	SANTA CLAUS	

Town famous for its Christmas postmark — **SANTA CLAUS**
Famous book by Lew Wallace — **BEN-HUR**
Most valuable farm product — **CORN**
Brothers who made cars in 1902 — **STUDEBAKER**
Famous automobile race held on ____ Day — **MEMORIAL**
One of Indiana's greatest natural resources — **SOIL**
Experimental community begun in 1825 — **NEW HARMONY**
Last name of two related U.S. Presidents — **HARRISON**

The circled letters spell the name of the Shawnee Indian chief defeated at the Battle of Tippecanoe. **Tecumseh**

| CAN YOU FIND... | What city was begun by the United States Steel Corporation in 1906? **Gary** |

Iowa
the Hawkeye State

Name _____

Iowa, the 29th state, is a farm state known as "the land where the tall corn grows." Its fertile soil makes the state a leading producer of corn, soybeans, beef cattle, hogs and dairy products.

The state's most important industries are related to agriculture, such as the manufacture of farm machinery and the processing of food products.

Iowa leads the nation in literacy — nearly everyone can read and write. Its school system has produced such famous people as President Herbert Hoover, Vice President (under President F. D. Roosevelt) Henry Wallace and artist Grant Wood.

MORE TO LEARN

Use the pictures on the map and the clues below to identify the towns, cities, bordering states and points of interest. Write the numbers in the △'s on the map.

1. Des Moines, largest city and state capital
2. Sioux City, nation's largest popcorn-processing plant
3. West Branch, site of Herbert Hoover's birthplace
4. Cedar Rapids, nation's largest cereal mill
5. Newton, nation's largest home-laundry appliance factory
6. Dubuque, named after Iowa's first settler
7. Minnesota — north
8. Nebraska — southwest
9. Illinois — southeast
10. Missouri — south
11. Wisconsin — northeast
12. South Dakota — northwest

eastern goldfinch

wild rose

| CAN YOU FIND... | Why is Iowa called the Hawkeye State? **It honors Black Hawk, a famous Indian chief.** |

Kansas the Sunflower State

Name _____

Kansas, the 34th state, is a rolling plain located in the center of the original 48 states. Near Osborne is the point (Geodetic Center) used as the reference to make all U.S. government maps.

The economy is varied in Kansas. It is second in flour milling, with the largest ones in Hutchinson. Wichita plants lead the nation in production of civilian aircraft. Government is the second largest employer and is centered around Fort Leavenworth and Topeka, the state capital.

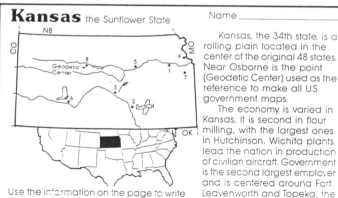

sunflower

western meadowlark

Use the information on the page to write the names of the places numbered on the map.

1. **Topeka** 2. **Wichita**
3. **Hutchinson** 4. **Ft. Leavenworth**
5. **Abilene** 6. **Dodge City**
7. **Osawatomie** 8. **Leavenworth**

MORE TO LEARN

Use the rhyming clues to fill in the blanks.

Kansas, an Indian word (m)eaning "swift w**ind**" (rhymes with mend)
A famous c(o)wboy town, D**odge** City (rhymes with lod(g)e)
Nation's l(e)ader in wh**eat** production (rhymes with ch(e)at)
Site of Dw(i)ght D. Eisenhower Library in Abilen(e) (rhymes with flight)
Abolitionist Joh(n) Br**own** State Park in Osawatomie (rhymes with crown)
S(a)nta Fe Tr**ail** established by Becknell in 1821 (rhymes with frail)

Unscramble the circled letters to complete the title of the state song "H**ome** on the R**ange**"

| CAN YOU FIND... | What gas does Kansas lead the world in production of? **helium** |

Answer Key

Kentucky
the Bluegrass State

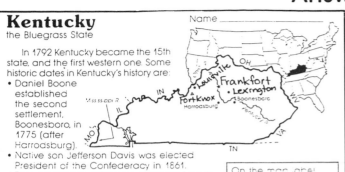

In 1792 Kentucky became the 15th state, and the first western one. Some historic dates in Kentucky's history are:
- Daniel Boone established the second settlement, Boonesboro, in 1775 (after Harroasburg).
- Native son Jefferson Davis was elected President of the Confederacy in 1861.
- The year before, native son Abraham Lincoln was elected U.S. President.
- The first Kentucky Derby was held in 1875, about 50 years after William McGuffey produced his first *McGuffey Reader*.
- The nation's gold vault was established at Fort Knox in 1936.

> On the map, label Frankfort, the capital, Fort Knox, Louisville and Lexington, the center of horse country.

MORE TO LEARN

Number the historic events listed below in the order they occurred.

5 Lincoln elected U.S. President
7 First Kentucky Derby held.
2 Boonesboro settled.
8 Nation's gold vault established.
1 Harrodsburg settled.
4 First *McGuffey Reader* produced.
3 Kentucky gained statehood.
6 Davis elected Confederate President.

Write the bordering states in alphabetical order. **Illinois, Indiana, Missouri, Ohio, Tennessee, Virginia, West Virginia**

CAN YOU FIND . . . What is located in Kentucky that is known as one of the Seven Natural Wonders of the modern world? **Mammoth Cave**

Page 17

Louisiana
the Pelican State

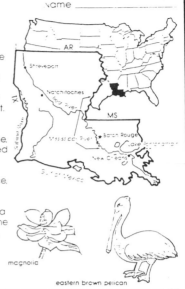

In 1812 Louisiana became the 18th state — the first one out of the original Louisiana Territory. Some historic dates in Louisiana's history are:
- LaSalle claimed the area for France in 1682 and named it.
- In 1718 de Bienville founded New Orleans.
- In 1760 the first Acadians came, two years before France ceded the land to Spain.
- In 1803 the U.S. bought the Louisiana Territory from France, three years after Spain had ceded it back to them.
- Two years before becoming a state, the eastern lands became the Republic of West Florida.
- In 1901 oil was discovered, forty years after Louisiana joined the Confederacy.

magnolia

eastern brown pelican

MORE TO LEARN

Number the historic events in the order they occurred.

2 New Orleans founded.
8 Louisiana became a state.
6 U. S. bought the Territory.
7 Republic of West Florida established.
1 Area named Louisiana.
10 Oil was discovered.
9 Louisiana joined the Confederacy.
3 Acadians first came.
4 France ceded area to Spain.
5 Spain ceded area to France.

CAN YOU FIND . . . What is a county in Louisiana called? **a parish**

Page 18

Maine
the Pine Tree State

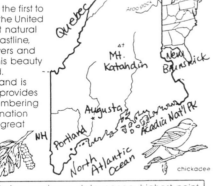

Maine, the 23rd state, is the first to greet the sun each day in the United States. It is a land of great natural beauty with its rugged coastline, sparkling lakes, rushing rivers and magnificent mountains. This beauty attracts tourists year round.

Ninety percent of the land is covered by woods which provides the product for its huge lumbering industry. Maine leads the nation with its lobster catch — a great delicacy in this country.

white pine cone and tassel

chickadee

> Label the capital, bordering provinces, state, ocean, highest point, national park and Portland (the largest city).

MORE TO LEARN

Write the number of the phase in Column B that describes who or what each is in Column A. Think logically!

A	B
8 Margaret Chase Smith	1. State capital
6 Mt. Katahdin	2. Only bordering state
7 Henry Wadsworth Longfellow	3. Backbone of the economy
5 Quebec and New Brunswick	4. Potato-producing county
2 New Hampshire	5. Bordering Canadian provinces
1 Augusta	6. Highest point on Atlantic Coast
10 Acadia National Park	7. Famous poet
3 Wood-processing industry	8. Both a U. S. Senator and Congresswoman
4 Aroostook	9. Carter's Secretary of State
9 Edmund Muskie	10. New England's only national park

CAN YOU FIND . . . Name a famous explorer from Maine. **Adm. Robert Peary**

Page 19

Maryland   the Old Line State

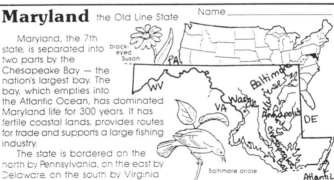

Maryland, the 7th state, is separated into two parts by the Chesapeake Bay — the nation's largest bay. The bay, which empties into the Atlantic Ocean, has dominated Maryland life for 300 years. It has fertile coastal lands, provides routes for trade and supports a large fishing industry.

The state is bordered on the north by Pennsylvania, on the east by Delaware, on the south by Virginia and on the west by West Virginia. Most of the population is centered around Washington, D. C. and Baltimore.

black-eyed Susan

Baltimore oriole

> Label the capital, largest city, nation's capital, bordering states, bay and ocean.

MORE TO LEARN

Using the WORD BANK, fill in the blanks to learn more about Maryland people and events. The circled letters, when unscrambled, spell the last name of the leader of the Underground Railroad during the Civil War, Harriet **Tubman**.

WORD BANK		
Francis Scott Key	District of Columbia	Baltimore
Spiro Agnew	Annapolis	Mason Dixon Line

Annapolis, state capital and home of the U. S. Naval Academy
District of Columbia, former Maryland territory
Francis Scott Key, writer of "The Star-Spangled Banner"
Baltimore, largest city in the state
Mason Dixon Line, state's northern boundary
Spiro Agnew, U. S. Vice President who resigned

CAN YOU FIND . . . What is the name of the fort that survived the British bombardment and inspired the writing of our national anthem? **Fort Mc Henry**

Page 20

Answer Key

Massachusetts
the Bay State

Name _____

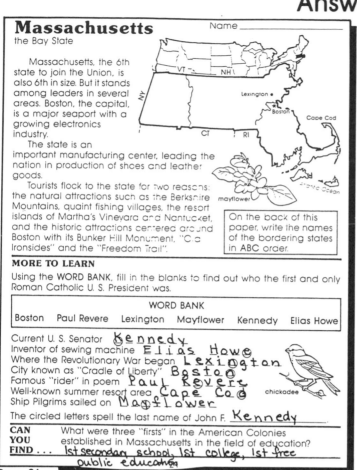

Massachusetts, the 6th state to join the Union, is also 6th in size. But it stands among leaders in several areas. Boston, the capital, is a major seaport with a growing electronics industry.

The state is an important manufacturing center, leading the nation in production of shoes and leather goods.

Tourists flock to the state for two reasons: the natural attractions such as the Berkshire Mountains, quaint fishing villages, the resort islands of Martha's Vineyard and Nantucket, and the historic attractions centered around Boston with its Bunker Hill Monument, "Old Ironsides" and the "Freedom Trail".

mayflower

On the back of this paper, write the names of the bordering states in ABC order.

MORE TO LEARN

Using the WORD BANK, fill in the blanks to find out who the first and only Roman Catholic U. S. President was.

WORD BANK					
Boston	Paul Revere	Lexington	Mayflower	Kennedy	Elias Howe

Current U. S. Senator **Kennedy**
Inventor of sewing machine **Elias Howe**
Where the Revolutionary War began **Lexington**
City known as "Cradle of Liberty" **Boston**
Famous "rider" in poem **Paul Revere**
Well-known summer resort area **Cape Cod**
Ship Pilgrims sailed on **Mayflower**

chickadee

The circled letters spell the last name of John F. **Kennedy**

CAN YOU FIND . . . What were three "firsts" in the American Colonies established in Massachusetts in the field of education? **1st secondary school, 1st college, 1st free public education**

Page 21

Michigan the Wolverine State

Name _____

Michigan, the 26th state, is often called the Great Lake State because it touches four of them: Lake Superior to the north, Lake Michigan to the west, Lake Huron to the northeast and Lake Erie to the southeast.

Some points of interest are:
- Detroit, the largest city and center of the auto industry
- Battle Creek, the world's largest producer of breakfast cereal
- Holland, site of the only authentic Dutch windmill in the nation
- Grand Rapids, former home of Gerald Ford and the site of his museum
- Lansing, the state capital
- Sleeping Bear Dunes National Lakeshore

The state's only national park, Isle Royale, has one of the largest herds of great-antlered moose left in the U.S.

apple blossom

On the map, label the points of interest and write them below.

★ **Lansing**
• **Detroit**   • **Holland**
• **Battle Creek**   • **Grand Rapids**
③ **Lake Superior** ③ **Lake Huron**
④ **Lake Erie**   ④ **Lake Michigan**
▲ **Isle Royale**   ▲ **Sleeping Bear Dunes**

robin

MORE TO LEARN

C	A	D	I	L	L	A	C	P
B	X	O	N	P	R	T	A	O
F	O	R	D	A	V	V	N	N
R	L	C	I	T	I	O	A	T
S	D	A	A	B	N	H	D	I
T	S	R	N	O	C	I	A	A
O	N	T	A	R	I	O	C	C
C	H	R	Y	S	L	E	R	V

Locate the following in the WORD SEARCH:
- The names of five makes of cars named after famous men in Michigan
- The only Great Lake not bordering Michigan
- The country bordering Michigan
- The two states on Michigan's southern border

CAN YOU FIND . . . Who founded the city of Detroit? **Cadillac**

Page 22

Minnesota the Gopher State

Name _____

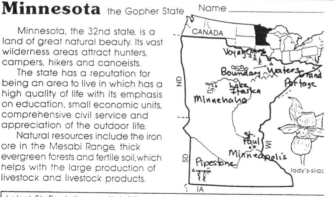

Minnesota, the 32nd state, is a land of great natural beauty. Its vast wilderness areas attract hunters, campers, hikers and canoeists.

The state has a reputation for being an area to live in which has a high quality of life with its emphasis on education, small economic units, comprehensive civil service and appreciation of the outdoor life.

Natural resources include the iron ore in the Mesabi Range, thick evergreen forests and fertile soil, which helps with the large production of livestock and livestock products.

lady's-slipper

Label St. Paul, the capital, Minneapolis, the largest city, national park and monuments, falls and wilderness.

MORE TO LEARN

Unscramble the letters to learn some interesting facts about Minnesota history, products and natural resources.

Leads the nation in production of these l o f u r (2 3 1 4 5) and n e n c d a (3 5 4 1 6 2) g e v e b a t s e l (3 4 1 2 7 6 5 10 9 8) **flour** and **canned vegetables**

Two national monuments a G r d n (1 2 5 4) r o P g e t a (3 2 1 6 7 4 5) and i s t P e p n e o (2 5 6 1 4 3 8 9 7) **Grand Portage** and **Pipestone**

Source of the Mississippi River k e L a (3 4 1 2) t a I s a k (2 3 1 4 6 5) **Lake Itaska**

Nation's only wilderness preserved for canoeists o B d u r a y n (2 1 5 3 7 6 8 4) a W t s r e (2 1 3 6 5 4) **Boundary Waters** Canoe Area

Famous falls in Longfellow's "Song of Hiawatha" n h n e M i a a h (3 6 4 5 1 2 9 7 8) **Minnehaha**

National park o g e V y u r s a (2 5 6 1 3 7 8 9 4) **Voyageurs**

loon

CAN YOU FIND . . . Who was the legendary lumberjack in Minnesota? **Paul Bunyan**

Page 23

Mississippi
the Magnolia State

Name _____

Mississippi, the 20th state, brings to mind pictures of the Old South — cotton plantations, stately magnolia trees and young men in gray.

Today only 25% of the goods produced are agriculturally based, with soybeans the most valuable crop. Manufacturing accounts for 66% with shipbuilding being the leading single industry.

Its cities include: Jackson, the capital and largest city; Vicksburg, site of crucial Civil War Battle; Natchez, site of classic antebellum mansions; Biloxi, Gulfcoast resort area; and Tupelo, Elvis Presley's birthplace.

magnolia

Label the cities on the map.

MORE TO LEARN

Mark each statement TRUE or FALSE. On another sheet of paper, rewrite each FALSE statement so it is TRUE.

F Mississippi is bordered by five states.
T Natchez is the site of many antebellum mansions.
F The state has no industry.
T Louisiana borders on the south.
F Biloxi is a well-known industrial center.
F Elvis Presley was born in Memphis, Tennessee.
T Jackson is the capital and largest city.
F Cotton is the most valuable crop.
T Shipbuilding is the leading single industry.
F Mississippians fought in the Union army.

mockingbird

CAN YOU FIND . . . What was the name of the last home of Jefferson Davis, President of the Confederacy? **Beauvoir**

Page 24

Missouri
the Show Me State

Name _____

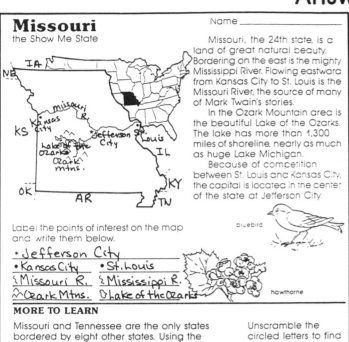

Missouri, the 24th state, is a land of great natural beauty. Bordering on the east is the mighty Mississippi River. Flowing eastward from Kansas City to St. Louis is the Missouri River, the source of many of Mark Twain's stories.

In the Ozark Mountain area is the beautiful Lake of the Ozarks. The lake has more than 1,300 miles of shoreline, nearly as much as huge Lake Michigan.

Because of competition between St. Louis and Kansas City, the capital is located in the center of the state at Jefferson City.

Label the points of interest on the map and write them below.

* Jefferson City
* Kansas City
* St. Louis
* Missouri R.
* Mississippi R.
* Ozark Mtns.
* Lake of the Ozarks

bluebird

hawthorne

MORE TO LEARN

Missouri and Tennessee are the only states bordered by eight other states. Using the directional clues below, label the bordering states on the map. Use postal abbreviations.

Iowa—(n)orth
Kansas—west
Illinois—e(a)st
Kentucky—so(u)theast
Tennessee—the (m)ost southeast

Arkansas—sou(t)h
Oklahoma—southwest
Nebraska—no(r)thwest

Unscramble the circled letters to find out the last name of the only U.S. President from Missouri.

Truman

| CAN YOU FIND . . . | Who were the French explorers that discovered the area which is Missouri? Joliet   Marquette |

Page 25

Montana
the Treasure State

Name _____

western meadow lark

CANADA
* Glacier National Park
Continental Divide
* Helena
* Butte   Billings
Custer Battlefield Nat'l. Mon.
WY

Montana, the 41st state, got its name from a Latin word meaning "mountainous". Located in the mountains are rich mineral deposits. The discovery of gold brought the first settlers — long after Lewis and Clark explored the region in 1805. Later, the copper deposits became more important. Today, petroleum is the state's leading mineral product. However, agriculture is one of the biggest money-makers, with large ranches and farms.

Its Glacier National Park, which borders Canada, has mountains never climbed and, perhaps, treasures to be discovered.

Fill in the blanks below using the symbols as clues.

* Helena   · Billings
· Butte   ✦ Glacier Nat'l Pk
--- Continental Divide
▲ Custer Battlefield Nat'l Mon.

bitterroot

MORE TO LEARN

Country bordering Montana	C A N A D A
Large economic moneymaker	A G R I C U L T U R E
First American explorers in the region	L E W I S A N D C L A R K
Montana means . . .	M O U N T A I N O U S
State capital	H E L E N A
Minerals which influenced state's early history	C O P P E R A N D G O L D

The circled letters spell the last name of the colonel who led the battle at Little Bighorn. Custer

| CAN YOU FIND . . . | Who was the Montana man who was a legend in Congress for thirty-four years? Mike Mansfield |

Page 26

Nebraska
the Cornhusker State

Name _____

SD
Niobrara River
IA
WY
Platte River
CO
Omaha
Lincoln ✶
MO
KS
goldenrod

Nebraska, which became the 37th state in 1867, is known for the hardy, pioneer spirit of the people.

The first "homestead" in the nation was claimed in 1862, ten years before J. Sterling Morton founded Arbor Day. When the area became the Nebraska Territory in 1854, it was nearly treeless. Now it has the only national forest in the nation planted by foresters.

On another sheet of paper, list the bordering states in ABC order.

In 1937 Nebraska adopted the only unicameral (one house) legislature in the nation, and in 1913 Leslie King, Jr. was born. Today he is known as Gerald Ford!

MORE TO LEARN

Number the historic events listed below in the order they occurred.

1 Nebraska (b)ecame a territory.
6 A (u)nicamera(l) (l)egislature was adopted.
4 J. Sterl(i)ng Morton (f)ounded Arbor Day.
3 Ne(b)raska became the 37th state.
5 Gerald Ford was born.
2 The (f)irst "h(o)mestead" w(a)s c(l)aimed.

western meadowlark

Write the circled letters. bullifbfoal When unscrambled, they spell the name of the man who organized a famous Wild West Show.

Buffalo Bill

| CAN YOU FIND . . . | What archeological find was unearthed near North Platte in 1922? largest mammoth fossil ever found |

Page 27

Nevada
the Silver State

Name _____

sagebrush
OR   ID
Reno
Great
* Carson City
Lake Tahoe
Basin
UT
Valley of Fire State Park
Las Vegas
AZ
Hoover Dam
mountain bluebird

Nevada, the 36th state, is a land of rugged beauty with its snow-capped mountains, sandy deserts and jagged plateaus.

Less rain falls in Nevada than in any other state. What brought people to the state was the rich mineral deposits — miners who wanted to strike it rich. Today 20 million tourists came to strike it rich — in the gambling casinos. Eighty percent of the state's residents live in or near Las Vegas and Reno and are dependent on the gambling industry for their livelihood.

MORE TO LEARN

Write the number of the phase in Column B that describes who or what each is in Column A. Think logically!

A		B
4	Comstock Lode	1. Formed Lake Meade, a huge man made lake
7	Tourism	2. Large desert area
3	Valley of Fire	3. State park with unusual rock formations
1	Hoover Dam	4. In 1859 a rich deposit of gold and silver
9	Lake Tahoe	5. State's largest city and gambling center
5	Las Vegas	6. President when Nevada admitted to the Union
6	Abraham Lincoln	7. Biggest industry in the state
8	Carson City	8. State capital
2	Great Basin	9. Beautiful lake resort area

| CAN YOU FIND . . . | What was the number of residents required for a territory to become a state in 1864? 127,381 How many residents were in Nevada in 1880? 62,266 |

Page 28

Answer Key

New Hampshire
the Granite State

Name _____

New Hampshire, the 9th state, is well known for its natural beauty. It is entirely bordered by Vermont on the west, by Massachusetts on the south, Canada on the north and Maine (and the Atlantic Ocean) on the east. Its Mt. Washington is New England's highest peak.

Though not important in New Hampshire's economy, its granite was used to build the Library of Congress and to make the cornerstone for the United Nations building.

A variety of people from the state played important roles in U.S. history: Franklin Pierce, the 14th President; Mary Baker Eddy, founder of the Christian Science movement; and Alan Shepard, first American astronaut in space.

Map labels: Canada, ME, purple finch, VT, Mt Washington, Concord, MA, Atlantic Ocean, purple lilac

Label the bordering states, country, ocean and the highest peak.

MORE TO LEARN
Mark each statement (T) TRUE or (F) FALSE. On another sheet of paper, rewrite each FALSE statement so it is TRUE.

__F__ Granite is important to New Hampshire's economy.
__T__ Mt. Washington is New England's highest peak.
__F__ Maine is bordered by only one state — New Hampshire.
__F__ Franklin Pierce was the 9th U.S. President.
__F__ Vermont borders the state on the east.
__F__ Alan Shepard was the first American to walk in space.
__T__ New Hampshire granite was used to build the Library of Congress.
__T__ Mary Baker Eddy founded the Christian Scientist movement.
__F__ New Hampshire was the 8th state to ratify the Constitution.

CAN YOU FIND . . . Who was the famous orator and senator in the 1800's? _Daniel Webster_

Page 29

New Jersey the Garden State

Name _____

New Jersey, the 3rd state, has been greatly influenced by its location. It lies between the Hudson and Delaware Rivers and along the Atlantic Coast. Products made there can be shipped throughout the U.S. and other countries.

Located between two giant cities, New York City and Philadelphia, the state has a huge nearby market for its farm products. Thousands of New Jerseyans work in these two cities and commute.

Located on the coast are many resort areas. Atlantic City is the most famous.

Map labels: eastern goldfinch, NY, PA, Paterson, Newark, Hudson River, Princeton, Trenton, Delaware River, Camden, Atlantic Ocean, Atlantic City, DE, purple violet

On another sheet of paper, list the major cities in ABC order.

MORE TO LEARN
Unscramble the letters to learn about some interesting people and events in New Jersey history.

asThmo aESion — **Thomas Edison** invented the electric light in his lab at Menlo Park.

maulSe oMres — **Samuel Morse** developed the electric telegraph near Morristown.

vGorer levCendal — **Grover Cleveland** 22nd U.S. President

doWorow lsnWio — **Woodrow Wilson**, 28th U.S. President

boHokne — **Hoboken**, site of the first pro baseball game in 1846

rteTonn rciePnnot — **Trenton** and **Princeton**, cities that served as the nation's capital

CAN YOU FIND . . . Where and when was the first dinosaur skeleton found in North America? _In Haddonfield in 1958_

Page 30

New Mexico
the Land of Enchantment

Name _____

New Mexico, the 47th state, is rich in history, beauty and natural resources. Founded in 1610 by the Spanish, Santa Fe is the oldest seat of government in the nation, and El Camino Real is the oldest road.

The Indiana influence can be seen in such names as the Navajo and Elephant Butte Dams and at such events as the Inter-Tribal Indian Ceremonial held each year.

A land of contrasts, New Mexico boasts the nation's greatest source of uranium ore (the first atomic bomb was exploded near Alamogordo in 1942), the beautiful White Sands desert area and Carlsbad Caverns, with the world's largest underground room.

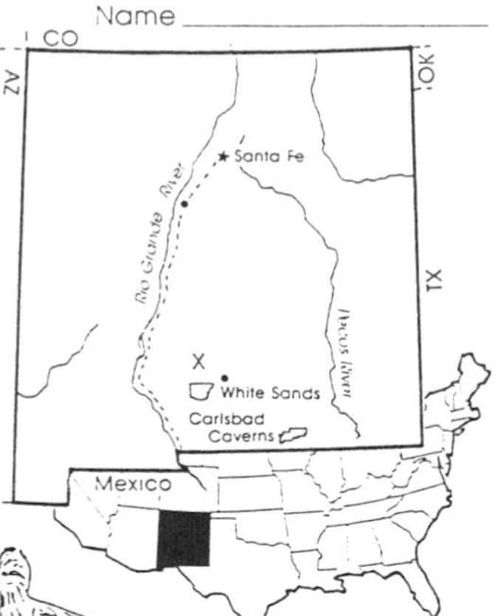

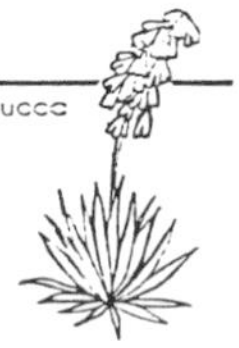

Map labels: CO, AZ, OK, Santa Fe, Rio Grande River, Pecos River, TX, X, White Sands, Carlsbad Caverns, Mexico, roadrunner, yucca

MORE TO LEARN

Navajo Dam created new farmland.
Elephant Butte Dam forms the state's largest lake.
White Sands, a large deposit of gypsum sand and a national monument
M ineral contributing to atomic research development, **uranium**
E l **Camino Real**, nation's oldest road
X marks the spot near **Alamogordo** of the first atomic bomb explosion
I ndian influence can be seen throughout the state.
C arlsbad Caverns one of the world's great natural wonders
O ldest seat of government in the nation, **Santa Fe**

CAN YOU FIND . . . Who found uranium in 1950 in the northwest region of New Mexico? _Paddy Martinez, a Navajo Indian_

Page 31

New York
the Empire State

Name _____

New York, the 11th state, has earned the nickname which came from a remark by George Washington predicting it might become the "seat of the new empire". And in a way, it has.

New York leads the nation in foreign trade and wholesale trade and is second in retail trade, manufacturing and population.

There is New York City, the nation's largest city, and center for banking, trade, communication, finance and transportation. As headquarters for the United Nations, it could be called the "capital of the world"!

Map labels: Lake Erie, Canada, VT, Lake Ontario, Niagara Falls, Buffalo, Cooperstown, Albany, MA, PA, NJ, CT, New York, bluebird, rose

Starting with Canada to the north, moving clockwise, label Vermont, Massachusetts, Connecticut, New Jersey, Pennsylvania, Lakes Erie and Ontario.
Label New York City, Albany (state capital), Cooperstown, Buffalo, Niagara Falls and the Erie Canal.

MORE TO LEARN
Use the WORD BANK to fill in the blanks.

WORD BANK			
Roosevelt	West Point	Statue of Liberty	Martin
Elmira	Millard Fillmore	Cooperstown	Van Buren

Location of the Baseball Hall of Fame **Cooperstown**
Last name of only cousins to be U.S. Presidents **Roosevelt**
U.S. Military Academy **West Point**
Location of Mark Twain's grave **Elmira**
Monument in New York Harbor **Statue of Liberty**
Two other U.S. Presidents from New York **Martin Van Buren** and **Millard Fillmore**

CAN YOU FIND . . . How much was paid to the Indians for Manhattan Island? _$24 in trinkets_

Page 32

North Carolina
the Tar Heel State

North Carolina, the 12th state to ratify the Constitution, leads the nation in tobacco farming and production of tobacco products. It also makes more wooden furniture and more cloth than any other state.

There are some beautiful, scenic areas in the Blue Ridge Mountain region which includes the Great Smoky Mountains and some lovely waterfalls. Pinehurst is a famous winter resort area in the western coastal plain

MORE TO LEARN

Number the cities, points of interest, bordering states and bordering bodies of water shown on the map.

5 Wright Brothers Monument at Kitty Hawk
3 Raleigh, the state capital
13 Great Smoky Mountains
6 Pamlico Sound
8 Fayetteville, on the Cape Fear River
4 Greensboro, world's largest mill for weaving denim
2 Kannapolis, world's producer of household textiles, such as towels
1 Mt. Mitchell, eastern American's highest peak
14 Charlotte, the largest city
7 Atlantic Ocean
12 Virginia, to the north
9 Tennessee, to the west
10 South Carolina, to the south
11 Georgia, to the southwest
15 Cape Hatteras, site of many shipwrecks

flowering dogwood

cardinal

CAN YOU FIND . . .	What was Blackbeard the Pirate's real name? Edward Teach

Page 33

North Dakota
the Flickertail State

North Dakota, the 39th state, has a higher percentage of people working in some form of agriculture than any other state. Only 15 cities have more than 2,500 people. The farmers have fought hard to keep cooperatives taking over from families.

Wheat is grown in every county. But North Dakota is also the nation's leader in production of barley, sunflower seeds and flaxseed.

Long ago the Sioux named the land "Dakota" meaning friends. That has modern significance in that the state built an International Peace Garden to commemorate the friendship between the U.S. and Canada.

MORE TO LEARN

Write the names of the following cities or points of interest.

* Bismarck * Fargo
‡ Theodore Roosevelt ⚡ Int'l Peace
  National Park   Garden
△ Garrison Dam •△ Rugby

western meadowlark

Write the number of the phrase in Column B that describes who or what each is in Column A. Think logically!

A
5 The only bank
7 North Central Rugby
8 –60° to 121°
1 Soil
2 International Peace Garden
3 Treaty with Sioux
4 Two SAC bases
6 Sacagawea

B
1. The state's most precious resource
2. Symbol of enduring friendship with Canada
3. Brought peace to the area
4. Located in North Dakota
5. Owned by a state
6. Famous Indian guide to Lewis and Clark
7. Geographic center of North America
8. U.S. record for one year temperature range

wild prairie rose

CAN YOU FIND . . .	Whose cabin is located on the grounds of the state capital? Theodore Roosevelt

Page 34

Ohio
the Buckeye State

Ohio, the 17th state, comes from the Indian word meaning "something great". The state has earned the name! Though 35th in size, it ranks 6th in population and 3rd in value of goods produced.

Some of the factors that have contributed to its growth and development are: its central location, its abundant water and fertile soils, and its valuable mineral resources.

Last, but not least, it was the home of seven U.S. Presidents!

scarlet carnation

Label the capital, cities and bordering states: Michigan — north, Indiana — west, Pennsylvania — northeast, West Virginia — southeast, Kentucky — south.

MORE TO LEARN

Circle the locations and some of Ohio's famous people in the WORD SEARCH.

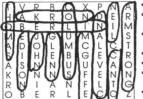

- Thomas Edison, inventor of the lightbulb
- John Glenn, U.S. Senator and former astronaut
- Canton, location of Pro Football Hall of Fame
- Oberlin, first coed college in the U.S.
- McGuffey, author of famous old readers
- Neil Armstrong, first astronaut on the moon
- Columbus, the capital
- Akron, world's greatest rubber producer
- Cleveland, largest city (near northern border)

CAN YOU FIND . . .	Who were the seven U.S. Presidents born in Ohio? Grant, Hayes, Benjamin Harrison, Taft, Garfield, McKinley, Harding

Page 35

Oklahoma the Sooner State

Oklahoma is an Indian word meaning "red people". In the 1800's the land was set aside as a huge Indian reservation for "as long as grass shall grow and rivers run". Today only about 4 percent of the population is Indian.

The Five Civilized Tribes wanted to become the state of Sequoyah in 1905. But Congress refused. So in 1907, the Oklahoma and Indian Territories united to become the 46th state.

The "Dust Bowl" of the 1930's is fertile land again for farming and raising cattle, and the discovery of oil brought a new source of wealth.

Beginning with Colorado, and moving clockwise, label the bordering states: Kansas, Missouri, Arkansas, Texas and New Mexico.

scissor-tailed flycatcher

MORE TO LEARN

Use the WORD BANK to fill in the blanks below.

Site of Ft. Sill, the Army's main artillery school Lawton
Famous humorist, the late Will Rogers
Choctaw chief who named the territory Allen Wright
What the Indian journey is often called Trail of Tears
A Cherokee Confederate brigadier general Stand Watie
The first capital of Oklahoma Guthrie
The present capital of Oklahoma Oklahoma City
Cherokee leader who invented a writing system Sequoyah

mistletoe

WORD BANK			
Sequoyah	Allen Wright	Trail of Tears	Guthrie
Oklahoma City	Will Rogers	Stand Watie	Lawton

CAN YOU FIND . . .	Why is Oklahoma called the "Sooner State"? Because of the settlers who "jumped the gun" to claim the best Indian land.

Page 36

Answer Key

Oregon the Beaver State

Oregon, the 33rd state, is known for its huge evergreen forests and rugged natural beauty. The mighty Columbia River on the north separates Oregon from Washington. The winding Snake River forms much of its boundary with Idaho on the east. Its beautiful shoreline on the Pacific Ocean is breathtaking. California and Nevada border on the south.

Portland, nicknamed the "City of Roses", is the largest city. Eugene ranks second. Salem is the capital. Crater Lake, Hells Canyon and the Columbia River Gorge help attract millions of tourists each year.

Name _____

Pacific Ocean · Portland · Columbia R. · WA · Columbia River Gorge · Salem · Eugene · ID · Crater Lake · CA · NV
Oregon grape

Label the bordering states, bordering bodies of water, cities and points of interest on the map. (13 items)

western meadowlark

MORE TO LEARN

Use the rhyming clues to fill in the blanks.

Oregon's most valuable crop, **wheat** (rhymes with cheat)
Resource that is one of the state's most valuable, **water** (rhymes with daughter)
Explorer who named Columbia River, Robert **Gray** (rhymes with pray)
Great, year-round skiing at Mount **Hood** (rhymes with good)
Oregon's most valuable industry, **Wood** processing (rhymes with hood)
Nation's deepest lake, **Crater** Lake (rhymes with greater)
The underlined letters, written in order, spell the name of the "Father of Oregon", John **McLoughlin**

CAN YOU FIND . . . What U.S. President based his campaign slogan (Fifty-Four Forty or Fight!) on the Oregon boundary dispute with the British? **James K. Polk**

Page 37

Pennsylvania
the Keystone State

Pennsylvania, the 2nd state to ratify the Constitution, was nicknamed "the Keystone State" because it was the center, or keystone, of the arch of the original 13 colonies. The Declaration of Independence was signed in Philadelphia.

The state today is experiencing difficulties in overcoming its dependence on three troubled industries — coal mining, steel production and railroading. Plus it has had to overcome the disaster in 1979 at the Three Mile Island Nuclear Power Plant. But Pennsylvania is in an era of renewal!

ruffed grouse · Name _____ · Lake Erie · NY · OH · Pittsburg · Harrisburg · Hershey · Philadelphia · NJ · WV · MD · DE

Label the bordering states and Great Lake starting with New York on the north, moving clockwise: New Jersey, Delaware, Maryland, West Virginia, Ohio, Lake Erie. Label Harrisburg, the capital, Philadelphia and Pittsburgh on the Ohio River.

MORE TO LEARN

Write the number of the name in Column B with the phase it matches in Column A. Think logically!

A	B
3 World's largest chocolate factory	1. William Penn
8 Bordering Great Lake	2. Pittsburgh
4 Declaration of Independence author	3. Hershey
1 The founder of Pennsylvania	4. Thomas Jefferson
7 Scientist, writer, philosopher, inventor	5. England
6 Largest city, cultural and industrial center	6. Philadelphia
2 America's steel capital	7. Benjamin Franklin
5 Country where the Liberty Bell was made	8. Erie

mountain laurel

CAN YOU FIND . . . Where is the location for the "official" groundhog for Groundhog Day? **Punxsutawney, PA**

Page 38

Rhode Island
Little Rhody

Rhode Island, the 13th state, may be the tiniest (48 miles long and 37 miles wide), but this state has had many firsts. Founded by Roger Williams in 1636 in his quest for religious freedom, the state had the first Baptist Congregation and the first Jewish Synagogue in the nation.

Rhode Island was also the location of other American firsts: first dry goods store, first cotton mill, first power loom, first torpedo boat . . . and, Rhode Island ranks first in the world in the production of costume jewelry and sterling silver products.

Name _____ · MA · Blackstone R. · Providence · CT · Narragansett Bay · Newport · Block Island Sound · Block Island
violet · Rhode Island red

MORE TO LEARN

ACROSS
1. _____ Bay
5. State capital
7. Made 1st _____ boat
8. State bordering on north

DOWN
1. City in southeast
2. First dry _____ store
3. _____ in size
4. Founder, _____ Williams
6. _____ jewelry center

Crossword solution:
NARRAGANSETT
PROVIDENCE
TORPEDO
MASSACHUSETTS
(TINIEST, NEWPORT, GOODS, COSTUME)

CAN YOU FIND . . . Who is the man from Rhode Island who is the oldest person ever to serve in Congress? **Theodore Francis Green**

Page 39

South Carolina
the Palmetto State

South Carolina was the 8th state to ratify the Constitution and the first state to secede from the Union. The battle that began the Civil War was at Fort Sumter. Sherman burned the capital city, Columbia.

The state is bounded by North Carolina on the north, Georgia on the west and the Atlantic Ocean on the east. Charleston, in the coastal lowlands, is its most historic city with an Old South flavor. Greenville, in the northwest, is the third largest city, after Columbia and Charleston.

Name _____ · NC · Greenville · Columbia · GA · Charleston · Atlantic Ocean
Carolina wren

Label the capital, large cities, bordering states and ocean.

MORE TO LEARN

Fill in the blanks from the WORD BANK

WORD BANK
tobacco   Calhoun   Santee   Citadel
Myrtle Beach   Parris Island

Famous seaside resort **Myrtle Beach**
Vice President under Jackson, John **Calhoun**
U.S. Marine base **Parris Island**
Military school at Charleston **Citadel**
Leading cash crop **tobacco**
Large river in the state **Santee**

Carolina jessamine

The circled letters spell the last name of a Revolutionary War hero known as the Swamp Fox, Francis **Marion**.

CAN YOU FIND . . . Who was the first Republican governor in 100 years elected in 1974? **James B. Edwards**

Page 40

Answer Key

South Dakota
the Sunshine State (Also known as the Coyote State.)

South Dakota, the 40th state, is a "Land of Infinite Variety" with its rich farmlands east of the wide Missouri River and its deep canyons and Badlands west of the river. Located in the Black Hills is Mount Rushmore, with the heads of four Presidents carved in its granite, and Homestake, the largest gold mine in the Western Hemisphere.

Sioux Falls, the largest city, gives evidence to the great place Indians had in the history of the state's development. A large monument is being built to Chief Crazy Horse.

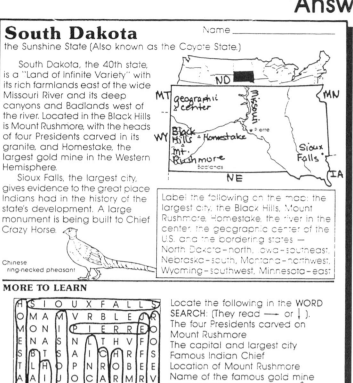

Label the following on the map: the largest city, the Black Hills, Mount Rushmore, Homestake, the river in the center, the geographic center of the U.S. and the bordering states —
North Dakota–north, Iowa–southeast, Nebraska–south, Montana–northwest, Wyoming–southwest, Minnesota–east

Chinese ring-necked pheasant

MORE TO LEARN

Locate the following in the WORD SEARCH: (They read → or ↓).
The four Presidents carved on Mount Rushmore
The capital and largest city
Famous Indian Chief
Location of Mount Rushmore
Name of the famous gold mine
Wide river in state's center

pasqueflower

CAN YOU FIND . . . What famous writer of children's books is from South Dakota?
Laura Ingalls Wilder

Page 41

Tennessee the Volunteer State

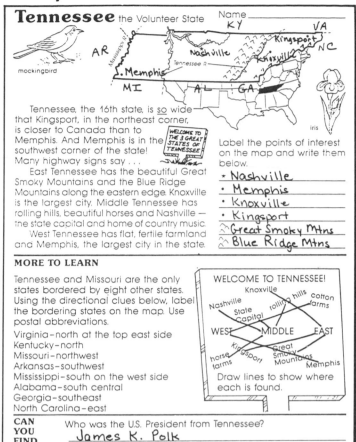

mockingbird

iris

Tennessee, the 16th state, is so wide that Kingsport, in the northeast corner, is closer to Canada than to Memphis. And Memphis is in the southwest corner of the state! Many highway signs say . . .

East Tennessee has the beautiful Great Smoky Mountains and the Blue Ridge Mountains along the eastern edge. Knoxville is the largest city. Middle Tennessee has rolling hills, beautiful horses and Nashville — the state capital and home of country music.

West Tennessee has flat, fertile farmland and Memphis, the largest city in the state.

Label the points of interest on the map and write them below.
• Nashville
• Memphis
• Knoxville
• Kingsport
⌢ Great Smoky Mtns
⌢ Blue Ridge Mtns

MORE TO LEARN

Tennessee and Missouri are the only states bordered by eight other states. Using the directional clues below, label the bordering states on the map. Use postal abbreviations.
Virginia–north at the top east side
Kentucky–north
Missouri–northwest
Arkansas–southwest
Mississippi–south on the west side
Alabama–south central
Georgia–southeast
North Carolina–east

WELCOME TO TENNESSEE!
Draw lines to show where each is found.

CAN YOU FIND . . . Who was the U.S. President from Tennessee?
James K. Polk

Page 42

Texas the Lone Star State

Texas, the 28th state, is big!! Big in size, big in population, big in products and big in history. Second only to Alaska in size, its King Ranch is bigger than Rhode Island!

The state has the most farms, farmland, cattle, horses and sheep in the nation. It leads the nation in the production of oil, natural gas and electrical power. It is the greatest U. S. source of salt, magnesium and sulphur.

The well-known cities are Houston, with the Manned Space Flight Center, Dallas, in the heart of the oil and cotton region, Ft. Worth, the cattle capital, and Austin, the capital.

mockingbird

MORE TO LEARN

Number the cities, points of interest, bordering states, country, rivers and body of water shown on the map.

8 Mexico, on the southwest
7 The famous Alamo in San Antonio
1 King Ranch
12 Manned Space Flight Center
13 Rio Grande River, bordering Mexico
15 Red River, bordering Oklahoma
14 New Mexico, on the west
18 Louisiana, on the southeast
11 Spindletop Monument, site of gusher oil

4 Ft. Worth
2 Dallas
3 Austin
6 San Antonio
10 Houston
16 Oklahoma, on the north
5 L. B. Johnson Home
17 Arkansas, on the northeast
9 Gulf of Mexico

bluebonnet

CAN YOU FIND . . . Who was the U. S. President assassinated in Dallas?
John F. Kennedy

Page 43

Utah
the Beehive State

"This is the place!" said Brigham Young in 1847 when he led Mormon settlers into the Salt Lake Valley to escape religious persecution. A year later the U.S. acquired the territory from Mexico. In 1850 it became the Utah Territory, a year after Young established the state of Deseret. Because of the Mormon practice of polygamy, it did not become the 45th state until 1896 (six years after suspending the practice). In fact, it even led to the Utah War in 1857.

The state bird is the seagull because, in 1848, these birds miraculously appeared and destroyed swarms of grasshoppers which were devouring the crops!

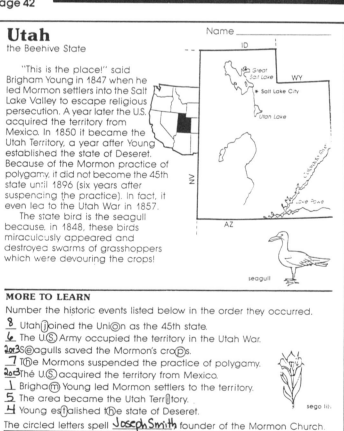

seagull

MORE TO LEARN

Number the historic events listed below in the order they occurred.

8 Utah joined the Union as the 45th state.
6 The U.S. Army occupied the territory in the Utah War.
2 or 3 Seagulls saved the Mormon's crops.
7 The Mormons suspended the practice of polygamy.
2 or 3 The U.S. acquired the territory from Mexico.
1 Brigham Young led Mormon settlers to the territory.
5 The area became the Utah Territory.
4 Young established the state of Deseret.

The circled letters spell Joseph Smith founder of the Mormon Church.

sego lily

CAN YOU FIND . . . What is the name of the ski resort developed by actor Robert Redford? Sundance

Page 44

Answer Key

Vermont
the Green Mountain State

Name _____

Vermont was the first state to join the newly-formed United States in 1791. The area had been awarded by the King of England to New York in 1764. But Ethan Allen and the Green Mountain Boys captured Fort Ticonderoga in 1775. An independent republic was established and a constitution adopted in 1777.

This constitution provided for state education, the right for every male to vote, and it forbid slavery. All firsts in America!

With its beautiful mountain scenery, Vermont is a great tourist area throughout the seasons.

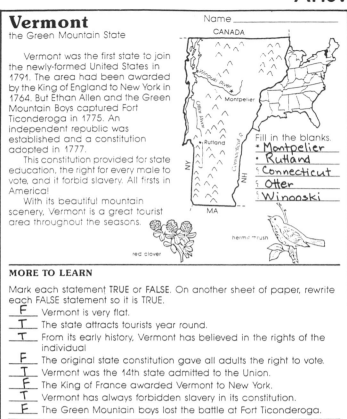

Fill in the blanks.
* Montpelier
* Rutland
* Connecticut
* Otter
* Winooski

MORE TO LEARN

Mark each statement TRUE or FALSE. On another sheet of paper, rewrite each FALSE statement so it is TRUE.

- **F** Vermont is very flat.
- **T** The state attracts tourists year round.
- **T** From its early history, Vermont has believed in the rights of the individual
- **F** The original state constitution gave all adults the right to vote.
- **T** Vermont was the 14th state admitted to the Union.
- **F** The King of France awarded Vermont to New York.
- **T** Vermont has always forbidden slavery in its constitution.
- **F** The Green Mountain boys lost the battle at Fort Ticonderoga.

CAN YOU FIND . . . What two U.S. Presidents were born in Vermont?
Chester A. Arthur and Calvin Coolidge

Page 45

Virginia Old Dominion

Name _____

Virginia, the 10th state, is more steeped in history than perhaps any other state. The list of important events is endless!

Nicknamed Old Dominion, Virginia is also known as "Mother of Presidents" as eight were born there and "Mother of States" as all or part of eight states were formed from it.

The greatest documents of freedom came from Virginians; the Declaration of Independence, the Constitution and the Bill of Rights.

Label the bordering states, bay and ocean, starting on the north and moving counterclockwise: Maryland, Chesapeake Bay, Atlantic Ocean, North Carolina, Tennessee, Kentucky and West Virginia.

MORE TO LEARN

Using the WORD BANK, fill in the blanks to learn about some historic events, places and people in our nation's history.

WORD BANK	Mt. Vernon	Pentagon	Williamsburg	Arlington
	Monticello	Appomattox	Jamestown	Thomas
	Patrick Henry	Robert E. Lee		Jefferson

Monticello, Thomas Jefferson's home
Mt. Vernon, George Washington's home
Appomattox, site of Confederate surrender
Jamestown, first permanent English settlement in America
Thomas Jefferson, author of the Declaration of Independence
Arlington, our National Cemetery
Williamsburg, restored colonial town
Patrick Henry, said, "Give me liberty or give me death"
Pentagon, the massive military center
Robert E. Lee, the outstanding Confederate general

CAN YOU FIND . . . Where and when did the British surrender to end the last major battle of the Revolutionary War? Yorktown, Oct. 19, 1781

Page 46

Washington
the Evergreen State

Name _____

Washington, the 42nd state, is a land of contrasts. The Cascade Mountains divide the state in two economic and geographic regions. In the east is one of the nation's most productive farm areas with its marketing center in Spokane. In the western lowlands are the industrial centers of Seattle and Tacoma. In the east are high mountains and thick forests. In the west are flat, treeless desert lands.

Water is one of its most important resources as it has more potential water power than any other state. The Grand Coulee Dam, the largest concrete dam in the U.S., is one of the world's greatest sources of water power.

Label the cities, the bordering country and the bordering states of Oregon and Idaho.

MORE TO LEARN

Mark each statement (T) TRUE or (F) FALSE. On another sheet of paper, rewrite the FALSE statements so they are TRUE.

- **F** All the land in Washington looks the same.
- **T** Mt. Rainier is the highest point.
- **T** The Grand Coulee Dam is the nation's largest concrete dam.
- **T** Olympia is the capital.
- **T** In the east are high mountains and thick forests.
- **T** Water is an important natural resource.
- **F** The Grand Coulee Dam is on the Snake River.
- **T** The industrial centers are in the western lowlands.

CAN YOU FIND . . . What volcanic mountain erupted in 1980?
Mt. Saint Helens

Page 47

West Virginia
the Mountain State

Name _____

West Virginia was born during the Civil War — as a result of the Civil War. It became the 35th state in 1863 because it had separated from Virginia when it seceded from the Union.

A rugged mountain land with no level ground and rocky soil, the people have had to look beneath it for a livelihood. Fortunately, there is an abundance of natural resources. — Coal is found under half the state, along with an abundance of natural gas, timber and salt deposits. And, tourism is developing as more visitors discover the beautiful mountains, forests, rivers and natural springs.

Label the bordering states starting with Pennsylvania to the north moving clockwise: Maryland, Virginia, Kentucky and Ohio.

MORE TO LEARN

Use the rhyming clues to fill in the blanks.
First stat(e) capit(a)l Wh**eeling** (rhymes with feeling)
Site (o)f John Brown's Raid, Ha(r)per's F**erry** (rhymes with warpers and merry)
(F)irst stat(e) governor, Arthur (B)**oreman** (rhymes with core and tan)
Diseas(e) common among coa(l) miners, bl**ack** l**ung** (rhymes with sack and sung)
Ran(k)s se(c)ond to Kentucky in p(r)oduction of c**oal** (rhymes with goal)
Highest point in the state, Sp**ruce** Kn**ob** (rhymes with truce and lob)
Write the circled letters. eloferelkcr
When unscrambled, they spell the name of the current U.S. Senator from a well-known family. Rockefeller

CAN YOU FIND . . . What town changed hands 56 times during the Civil War?
Romney

Page 48

Our 50 States IF8749   114

Answer Key

Wisconsin
the Badger State

Name _____

Wisconsin, the 30th state, is known as "America's Dairyland". It is the nation's leading producer of milk and cheese.

But manufacturing is the state's chief industry. It is a leader in manufacturing engines and turbines, in canning vegetables and brews more beer than any other state.

The natural beauty and 8000 plus lakes attract millions of vacationers. Wisconsin has won fame for its high quality of progressive government and education.

Label Madison, the capital, the other cities, and bordering states and Great Lakes starting with Lake Superior on the north and moving clockwise: Michigan, Lake Michigan, Illinois, Iowa, Minnesota and Lake Superior.

MORE TO LEARN

Write the number of the name in Column B that matches the phrase in Column A. Think logically!

	A		B
6	A bordering Great Lake	1.	Joseph McCarthy
3	Celebrated pro football team	2.	Winnebago
5	Most important type of farming	3.	Green Bay Packers
7	Brothers who started their circus there	4.	Milwaukee
8	Political party founded at Ripon	5.	Dairying
1	Controversial U.S. Senator elected in 1946		
4	The largest city (located on Lake Michigan)	6.	Superior
		7.	Ringling
2	State's largest lake	8.	Republican

robin

CAN YOU FIND . . . Who was the famous architect from Wisconsin?
Frank Lloyd Wright

Page 49

Wyoming
the Equality State

Indian paint brush

Name _____

Wyoming, the 44th state, is known for its beautiful mountains and parks. Millions of tourists visit the state each year.

The soil is poor, so 95 percent of the land is used for grazing. Cattle and oil are the foundation of the economy. Since half the land is federally-owned and controlled, the government plays an important role in Wyoming's future.

Cheyenne, the capital and largest city, has less than 50,000 people. The next largest cities are Laramie and Casper, on the North Platte River.

Label the capital, cities, national monument and parks, and bordering states starting with Montana on the north, moving clockwise: South Dakota, Nebraska, Colorado, Utah and Idaho.

MORE TO LEARN

Fill in the blanks using the WORD BANK.

WORD BANK	Devils Tower	Grand Tetons	Yellowstone
	Shoshone	Esther Morris	Sacajawea

meadow lark

The first national monument *Devils Tower*
The first, and largest, national park *Yellowstone*
The first national forest *Shoshone*
The nation's first woman justice of the peace *Esther Morris*
The Shoshone girl guide of Lewis and Clark *Sacajawea*
One of the world's most dramatic mountains *Grand Tetons*

Write the circled letters. *rweran* Unscramble them to find the name of the site of the first U.S. long range missile squadron.
Warren Air Force Base

CAN YOU FIND . . . What famous family helped the Grand Tetons become a national park? *Rockefeller*

Page 50

The Grand Canyon

Name _____

The Grand Canyon in Arizona was once covered by water. The earth's crust began to move over two billion years ago and raised itself one and a half miles above sea level. This mountain-building process and rushing water began to carve the canyon. The Colorado River started shaping the canyon about six billion years ago. As the river cut its way down, it exposed several different kinds of rock which make up the canyon's walls. The canyon is about one mile deep and is anywhere from 6 feet to 18 miles wide. The Colorado River flows through it for 277 miles. Visitors may view the canyon from several different points along the North and South Rims, or they may hike down on one of its many trails.

Besides its geologic history, the Grand Canyon has a history of man. Several Indian tribes have lived in it the past 4000 years. Now a small tribe called the Havasupai live at the bottom of the canyon. Spanish explorers were the first to see the canyon. John Wesley Powell, an American geologist, was the first to travel through it. He gave the "big canyon" its name.

Write True or False in front of the statements below.

F The Grand Canyon began to form two billion years ago.
F The Grand Canyon is in Colorado.
F Spanish explorers named the Grand Canyon.
F The canyon is 18 miles deep.
T There was once a sea covering the Grand Canyon.
F The canyon first began when the Colorado River cut through it.
T There are two rims around the Grand Canyon.
F John Wesley Powell was the first to see the Grand Canyon.
T The Grand Canyon is made of several kinds of rock.
F No one can live in the canyon.
F The only way to see the canyon is from the top.
F The Grand Canyon is named as it is because of its size.
F The formation of the canyon began when the earth's crust erupted.

Page 51

Niagara Falls

Name _____

Niagara Falls is in the Niagara River. The river forms part of the border between New York and Ontario, Canada. The river flows gently out of Lake Erie. Before it reaches Lake Ontario, it falls dramatically over Niagara Falls. Goat Island is in the middle of the Niagara River. The island divides the river into two falls. It sends eighty-five percent of the river over Horseshoe Falls on the Canadian side and fifteen percent of it over the American Falls on the United States side. Niagara Falls drops straight down into a deep gorge. The limestone rock ledge at the top is harder than the rock below it. The water has eroded the softer stone below and created the Cave of the Winds behind the American Falls. Pieces of the harder rock above have broken off occasionally, and over the years the gorge has become about seven miles longer. Visitors may look at the falls from Goat Island or on a sightseeing steamer on the river.

Answer the questions. Write the circled letters in order. They will spell the name of one of the sightseeing boats. Use the underlined words above to help find the answers.

What divides the two falls? *Goa(t) Island*
What is the name of the falls on the Canadian side? *Horsesho(e)*
What is the name of the falls on the United States side? *America(n)*
On what river are the falls? *Niag(a)ra*
In what country is Ontario? *Ca(n)ada*
Into what lake does the Niagara River flow? *(O)ntario*
What percent of the falls are on the United States side? *(F)ifteen*
What is behind the American Falls? *Cav(e) of th(e) Winds*
What kind of rock is on the ledge? *lim(e)s(t)one*
From what lake does the Niagara River flow? *Eri(e)*
On what kind of sightseeing boat may visitors view the falls? *(S)teamer*
The name of one boat is *The Maid of the Mist*

Page 52

Answer Key

The Great Plains

Name _____

The Great Plains extend 2500 miles north and south from Canada to Texas, and east and west from the Rocky Mountains 400 miles to the central lowlands. Texas, Oklahoma, Kansas, Nebraska, South Dakota, North Dakota, New Mexico, Colorado, Wyoming and Montana are part of the Great Plains. The western boundary is higher and slopes toward the eastern boundary at ten feet per mile. It is not noticeable when one looks out into the distance.

Pioneers crossed the Great Plains in wagons. It was a long trip with little to see but grasslands. It is much the same way today, but the highway is improved. There is still grassland where wheat does not grow on the farms that have been built. There are no bison or buffalo.

Fill in the boxes below with the names of the states that are part of the Great Plains. One letter has been put in to get you started. When you have written the states' names in the boxes, write them again alphabetically on the lines to the right.

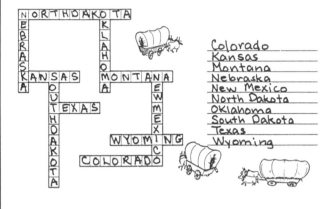

Colorado
Kansas
Montana
Nebraska
New Mexico
North Dakota
Oklahoma
South Dakota
Texas
Wyoming

Page 53

The Colorado Plateau

Name _____

The Colorado Plateau is a little larger than what is called the Four Corners Region. The area of the Colorado Plateau covers the northern third of Arizona, southern two-thirds of Utah, northwestern quarter of New Mexico and the southwestern quarter of Colorado. The plateau is flat with deep cut canyons. The Grand Canyon is the best known. The Colorado River cuts through several of the canyons; the Colorado National Monument, past the southern end of Arches National Park, Canyonlands National Park, Glen Canyon and through the Grand Canyon. Lake Powell was created when Glen Canyon Dam was built to hold back the Colorado River. Rainbow Bridge National Monument may be reached by boat on Lake Powell. The Colorado River was not the main force to shape some of the places in the region. Natural Bridges National Monument was cut by two rivers. Zion National Park was cut by the Virgin River. Cedar Breaks National Monument, Capitol Reef National Park and most of Arches National Park were created by rain, streams, wind and ice.

Alphabetize the names of the underlined places. If lined up correctly, each will form a sentence about itself.

Arches National Park — has a formation called Landscape Arch.
Bryce Canyon National Park — has fourteen valleys filled with imaginative shapes like monks and temples.
Canyonlands National Park — is cut by the Green River.
Capitol Reef National Park — has a white-capped sandstone ridge.
Cedar Breaks Nat'l Monument — has steep, pink, cliff formations.
Colorado Nat'l Monument — has a shape called Devil's Kitchen.
Glen Canyon — is a national recreation area.
Grand Canyon Nat'l Park — has one rim higher than the other.
Lake Powell — was created when Glen Canyon Dam was built.
Natural Bridges Nat'l Monument — has three natural sandstone bridges.
Rainbow Bridge Nat'l Monument — is the world's largest known natural bridge.
Zion National Park — has a rock mass called Checkerboard Mesa.

Page 54

Deserts

Name _____

There are other desert regions in the southwestern United States besides Death Valley and the Great Basin. The Sonoran desert in southwestern Arizona and southeastern California is not a deserted, sandy place as one might think. All sorts of animal and plant life live in it. Cactus is the main plant. Its flower provides food for birds. Woodpeckers make their homes in the cacti. Lizards and other ground animals often sleep during the day and hunt for their food at night. The Colorado, Yuma and Succulent Deserts are part of the Sonoran Desert. The Mojave Desert is in California between the Sierra Nevada mountains and the Colorado River. The Mojave Desert was once covered by the Pacific Ocean, but when the mountains rose they kept out the water from the sea. The Painted Desert is in northern Arizona. Buttes, mesas, pinnacles, valley formations and varied colors cover the area.

Circle the names of eight American deserts in the puzzle. Write them in alphabetical order on the lines to the right. The names read ⇄ | ↗ ↙

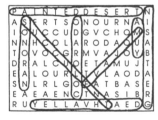

Colorado
Death Valley
Great Basin
Mojave
Painted Desert
Sonoran
Succulent
Yuma

Page 55

Indian Ruins

Name _____

The Four Corners region in the United States, where Arizona, Utah, Colorado and New Mexico meet, was occupied hundreds of years ago by the ancestors of today's American Indians. The remains of their dwellings, pottery and tools tell what their life was like about 800 years ago. The United States Government has set many of these sites aside as national monuments and parks to preserve their history and scenic beauty.

Names of some of these sites are scrambled below in a sentence about them. Unscramble the name and write it under the picture of its ruin.

N A M E U O M Z T   S C E T L A   National Monument
3 9 1 5 7 2 8 6 4   12 10 15 13 14 11   is a five story building built in a niche on the side of a cliff.

E S M A   R V E E D   National Park has many structures that
2 3 1 4   7 5 6 9 8   were built beneath an overhang in a cliff.

L E N R B I D A E   National Monument has the ruins of a meet-
6 5 3 1 9 7 4 2 8   ing house that looks something like a maze. This was done for defense.

V N O A A J   National Monument has three ancient cliff dwellings
3 1 6 2 4 5   and sits in the middle of the Navajo Indian Reservation today.

MONTEZUMA CASTLE

MESA VERDE

BANDELIER   NAVAJO

Page 56

Our 50 States IF8749   116   © 1992 Instructional Fair, Inc.

Answer Key

The Mississippi River

Name _____

Lake Itasca, MN, WI, IA, IL, St. Louis, MO, KY, TN, AR, MS, LA, New Orleans, Gulf of Mexico

Chippewa <u>Indians</u> gave the name Messipi to the Mississippi River which means great river. The Mississippi is the longest river in the United States. It is 2348 miles long. Its <u>source</u> is a <u>stream</u> flowing out of Lake <u>Itaska</u> in northwest Minnesota. Its beginning is so small you can wade across it. As it travels, thousands of streams and rivers join it to make it the large river it is. The Mississippi River forms all of part of the eastern borders for the states of <u>Minnesota</u>, <u>Iowa</u>, Missouri, Arkansas and Louisiana and the western borders for Wisconsin, <u>Illinois</u>, Kentucky, Tennessee and Mississippi. As the river travels, it moves faster and <u>picks up</u> and carries <u>sand</u>, mud and <u>pebbles</u>. This is deposited as <u>silt</u> along the lower banks and in the mouth of the river. The river ends its journey in the Gulf of Mexico.

There is a rhythm to spelling M·I·S·S·I·S·S·I·P·P·i that is fun. The answer to each question below begins with a letter of the river's name. Let the underlined words above help you.

In what state does the Mississippi River begin? — **M**innesota
What is one state it borders? — **I**owa
What is something it carries? — **S**and
What is something it deposits? — **S**ilt
What is the lake's name where it begins? — **I**taska
What does the river begin as? — **S**tream
What is another word that means beginning? — **S**ource
Who named it Messipi? — **I**ndians
What do it do with mud and sand? — **P**icks up
What else does it pick up? — **P**ebbles
What is another state it borders? — **I**llinois

Page 57

Tributaries of the Mississippi

Name _____

A tributary is a smaller stream or river that empties into a larger body of water. Every stream or river that flows must go somewhere. Almost half of the rivers in the United States empty into the Mississippi River. These rivers flow between the Rocky Mountains in the west and the Appalachian Mountains in the east. The Missouri is the second longest river. It flows into the Mississippi just above St. Louis, Missouri, after traveling 2,315 miles from its source in Montana, through South Dakota, Iowa, Kansas and Missouri. It was nicknamed Big Muddy by the pioneers because it was so dirty. The Arkansas River begins high in the Rockies in Colorado and flows through Kansas, Oklahoma and Arkansas before it empties into the Mississippi River. The Ohio River flows 981 miles through coal fields, steel factory districts and farm lands. It flows from Pennsylvania along the borders of West Virginia, Kentucky, Ohio, Indiana and Illinois. It empties into the Mississippi at its widest point—Cairo, Illinois.

Write True or False to the following statements.
___F___ The Missouri River is longer than the Mississippi.
___T___ The Arkansas River begins in Colorado.
___T___ The Ohio River is the smallest of the rivers mentioned.
___F___ The Mississippi is a tributary to these rivers.
___F___ The Ohio empties into the Mississippi just above St. Louis.
___F___ Pennsylvania is where the Missouri River begins.
___T___ The Missouri River goes through or by six states.
___F___ Over half the rivers from the Rockies to the Appalachians flow into the Mississippi.
___F___ The Ohio River flows through Oklahoma, Kansas and Arkansas.
___F___ The Missouri's widest point is in Cairo, Illinois.
___F___ Big Muddy is a good name for the Arkansas River.
___T___ The Arkansas River is larger than the Ohio River.
___T___ The Appalachian Mountains are in the east.
___T___ Sometimes a river can be a border between states.

Page 58

Rivers in the West

Name _____

There are several important rivers in the western United States that do not empty into the Mississippi River. They are best known for their beauty and recreational activities including rafting, water skiing and fishing. There is not as much water in the western United States so the rivers there have also been used in the building of dams, lakes and reservoirs.

Work the math problems below. The answers to the problems will spell the name of some of the rivers in the west. Use the CODE BOX on the left to decode each river's name. Write it out correctly and read a fact about that river.

CODE BOX

A = 26	N = 25
B = 24	O = 23
C = 22	P = 21
D = 20	Q = 19
E = 18	R = 17
F = 16	S = 15
G = 14	T = 13
H = 12	U = 11
I = 10	V = 9
J = 8	W = 7
K = 6	X = 5
L = 4	Y = 3
M = 2	Z = 1

$$\begin{array}{cccccccc} 30 & 32 & 16 & 5 & 11 & 13 & 28 & 17 \\ -8 & -9 & -12 & +6 & -9 & +11 & -18 & +9 \\ \hline 22 & 23 & 4 & 11 & 2 & 24 & 10 & 26 \\ C & O & L & U & M & B & I & A \end{array}$$

This river is the second longest river in the Western Hemisphere to empty into the Pacific Ocean.

$$\begin{array}{ccccc} 7 & 12 & 13 & 19 & 25 \\ +8 & +13 & +13 & -13 & -7 \\ \hline 15 & 25 & 26 & 6 & 18 \\ S & N & A & K & E \end{array}$$

This river is a tributary of the Columbia River.

$$\begin{array}{cccccccc} 15 & 42 & 11 & 14 & 9 & 15 & 14 & 17 \\ +7 & -19 & -7 & +9 & +8 & +11 & +6 & +6 \\ \hline 22 & 23 & 4 & 23 & 17 & 26 & 20 & 23 \\ C & O & L & O & R & A & D & O \end{array}$$

This river has carved the Grand Canyon for millions of years.

$$\begin{array}{ccccccccc} 23 & 3 & 12 & 7 & 28 & 19 & 14 & 36 & 29 \\ -6 & +7 & +11 & +7 & -11 & +7 & +11 & -16 & -11 \\ \hline 17 & 10 & 23 & 14 & 17 & 26 & 25 & 20 & 18 \\ R & I & O & G & R & A & N & D & E \end{array}$$

This river forms an international border between the United States and Mexico.

Page 59

The Great Lakes

Name _____

Duluth, Superior, CANADA, WI, Milwaukee, Michigan, Huron, Ontario, MN, MI, Detroit, NY, Chicago, IN, Erie, Cleveland, PA, IL, OH

The Great Lakes began to form over 250,000 years ago when large sheets of ice, called glaciers, dug holes in the land. Melting glacial snows filled the holes with water and formed the Great Lakes, the world's largest group of fresh water lakes. There are five lakes (in order of size): Superior, Huron, Michigan, Erie and Ontario. It is possible to travel on a boat from the Atlantic Ocean, through the Saint Lawrence Seaway, over Lakes Ontario, Erie, Huron and Michigan, onto the Illinois River, into the Mississippi River to the Gulf of Mexico.

Use the map and information above to help answer the questions below. Write the circled letters in your answers on the blanks at the bottom of the page. They will spell the name of the shallowest Great Lake.

Color Lake Superior blue, Lake Michigan green, Lake Huron orange, Lake Erie red and Lake Ontario purple.

Which lake is the largest? **S**uper**i**or
Which lake does not have Canada as one of its boundaries? **M**ichigan
How many lakes form the Great Lakes? fiv**e**
From what did the water come to fill the Great Lakes? glaciers
Which lakes touch New York? **E**rie and **O**ntario
Which state is touched by Lakes Michigan and Superior? W**i**sconsin
What Michigan city is an important port on the Great Lake system? D**e**troit
The shallowest Great Lake is Lake **E**rie.

Page 60

Answer Key

The Great Salt Lake and Its Surroundings

Name _____

The Great Salt Lake and Great Salt Lake Desert are part of the Great Basin. The Great Basin is a large desert area extending into parts of six states; California, Idaho, Oregon, Nevada, Utah and Wyoming. It is called a basin because all of its bodies of water stay within it. Sinks lie in some of its valleys. The Great Salt Lake is its largest sink.

The Great Salt Lake is in Utah. It is considered one of the natural wonders of the world. A fresh water lake was once in its place thousands of years ago, but it dried up leaving several small lakes. The Great Salt Lake is the largest of the remaining lakes. Today the lake gets its water from rain and fresh water streams, but its salt is from the salt deposits left by the original dried-up lake. When the lake is low, it has several islands. Cattle are raised on the largest one, Antelope Island. The smaller islands are breeding grounds for gulls, ducks, geese and pelicans. There are no fish in the lake, only shrimp. The Great Salt Lake Desert is a low, flat, dry region to the west of the lake. A part of it is so hard that it is used as a track for automobile racing.

Use the underlined words above to help answer the clues below. Write your answers on the blanks. The name geologists give the Great Salt Lake will appear down in the boxes.

Clue	Answer
Place in which water stays	B**a**s**i**n
State into which Great Basin extends	Idah**o** — Basin
Island where cattle raised	A**n**telope
What is in some valleys	s**i**nks
Birds that nest on smaller islands	g**e**ese
State into which Great Basin extends	Ne**v**ada
Lives in Great Salt Lake	shr**i**mp
Birds that nest on smaller islands	gu**l**ls
State into which Great Basin extends	Ca**l**ifornia
Birds that nest on smaller islands	p**e**licans
Geologists call it	Bonneville Lake.

Page 61

The Appalachian Mountains

Name _____

The Appalachian Mountains extend 1500 miles across the Eastern United States between Alabama and Canada. The Appalachian Mountains are the oldest mountain system in North America and the second largest system after the Rocky Mountains. All the main ranges east of the Mississippi River are in the Appalachians except for the Adirondacks. The ranges in the United States that are a part of this system are the White Mountains, Green Mountains, Catskill Mountains, Great Smoky Mountains, Allegheny Mountains, Cumberland Mountains and the Blue Ridge Mountains. Mount Washington is the tallest mountain in New Hampshire and has a weather station on top. Mount Mitchell, in the Blue Ridge Mountains, is the tallest mountain in the Appalachians. Spruce Knob is the tallest peak in the Allegheny range.

Write the scrambled answer to each clue below on the line following it. Let the underlined words above help you.

Where the Appalachian Mountains are located
ARSNEET **Eastern**
NIUETD **United**
TTSSAE **States**
Mountain range that is not part of the Appalachian system
DIAASRCODNK **Adirondacks**
The oldest mountain system in North America
ACAPAINAPLH **Appalachian**
Where there is a weather station
ONMUT **Mount**
NSITAGNWHO **Washington**
The mountains in which the tallest Appalachian peak is located
LEUB DEGRI **Blue Ridge**
The tallest peak in the Allegheny range
PCUSRE KOBN **Spruce Knob**
Write the mountain ranges in alphabetical order on the back of this paper.

Page 62

The Rocky Mountains

Name _____

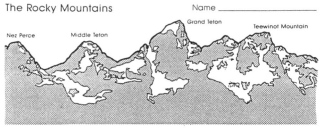

Nez Perce   Middle Teton   Grand Teton   Teewinot Mountain

The Rocky Mountain chain is the largest mountain system in North America. It extends more than 3000 miles through the United States and Canada. It is 350 miles wide in some places. In the United States the range extends through New Mexico, Colorado, Utah, Wyoming, Idaho, Montana and Alaska. The high peaks in the Rockies form the Continental Divide which separates the direction water flows—toward the Atlantic or Pacific Oceans. It crosses New Mexico, Colorado, Wyoming, Idaho and Montana in the United States. The Columbia, Missouri, Colorado, Arkansas and Rio Grande Rivers begin high in the Rockies. Yellowstone, Grand Teton and Rocky Mountain National Parks are in the Rockies as is Waterton/Glacier International Peace Park.

Circle as directed in the puzzle below. Some names will be circled two or three times.

1. The names of the states the Rocky Mountain system is in. Circle them red.
2. The names of the National Parks in the Rocky Mountains. Circle them green.
3. The names of the rivers that begin in the Rockies. Circle them blue.
4. The names of the states the Continental Divide crosses. Circle them yellow.

```
E G R A N D E O I W O L L E Y W
W R A N C H E R I O Y M R N O E
A A N A T N O M L O O A T M D O
I N D A R O L O C U M R I N A
R D N E R G T M I K G O T I G A
U T A H R I S Y T E N A S O N Y
O E O M A T W M N T I C E E G
S T T O R H H A T M B O N W O
S O R N A I L N R M U M L O M I
I N E A N A L A G R Y U A T E R
M I T T A M E C L W L M S X A
C H A U T E Y W A E A D B O I D
O I D A H O A R K S A O A L C O
N R U G A S N A K R A U I L O H
O O R I O O D A R O L O C E A D
M I S O U R R I C O L U M Y R I
```

Page 63

The Cascades

Name _____

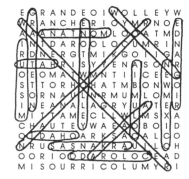

The Cascade Range is a chain of mountains that extend from northern California, through western Oregon and Washington, into British Columbia. The range is named after large cascades the Columbia River makes as it cuts through the mountains.

The mountains in this range were made from lava flows occurring over thousands of years. Most of the range's peaks are extinct volcanoes. Mount Rainier and Mount Adams in Washington, Mount Hood in Oregon and Mount Shasta in California are four inactive volcanoes. Crater Lake, in the southern part of the range, is all that remains of ancient Mount Mazama which erupted 7000 years ago. Lassen Peak in California was inactive for many years, but from 1914–1921 it had several eruptions. Mount Saint Helens in Washington had been quiet for over 100 years, when in 1980 it began to erupt again. More than 1000 feet of its peak has been blown off in its recent eruptions, and a large crater has developed. Scientists do not know when the next eruption will be, but they keep an eye on the Cascades.

Answer True or False to the following statements.

__T__ The Cascades were made from volcanic eruptions over thousands of years.
__T__ Most of the volcanoes in the Cascades are extinct.
__F__ Mount Saint Helens had never erupted before 1980.
__F__ Lassen Peak is in Oregon.
__T__ The Cascade Range extends from Northern California into Canada.
__T__ Crater Lake is a volcano.
__F__ Mount Saint Helens has become higher since 1980.
__T__ The Columbia River cuts through the Cascades.
__T__ Mount Rainier and Mount Hood are inactive volcanoes.
__F__ Lassen Peak has not erupted in over 100 years.
__F__ The Cascades never have any volcanic activity now.
__T__ Scientists are interested in the Cascades.
__T__ Mount Rainier and Mount Adams are in Washington.
__F__ From 1912–1941 Lassen Peak had several eruptions.
__T__ When a volcano is active it erupts.

Page 64

Coast Ranges

Name _____

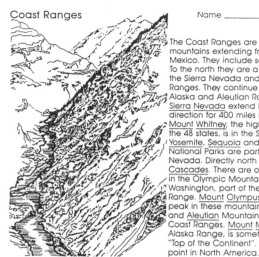

The Coast Ranges are a chain of mountains extending from Alaska to Mexico. They include several ranges. To the north they are a continuation of the Sierra Nevada and Cascade Ranges. They continue on as the Alaska and Aleutian Ranges. The <u>Sierra Nevada</u> extend in a north-south direction for 400 miles in California. <u>Mount Whitney</u>, the highest peak in the 48 states, is in the Sierra Nevada. <u>Yosemite</u>, <u>Sequoia</u> and <u>Kings Canyon</u> National Parks are part of the Sierra Nevada. Directly north are the <u>Cascades</u>. There are over 100 glaciers in the Olympic Mountains in northern Washington, part of the <u>Pacific Coast</u> Range. <u>Mount Olympus</u> is the highest peak in these mountains. The <u>Alaska</u> and <u>Aleutian</u> Mountains end the Coast Ranges. <u>Mount McKinley</u>, in the Alaska Range, is sometimes called "Top of the Continent". It is the highest point in North America.

The correct answers to the clues below will spell COAST RANGES down. Use the underlined words above to help you fill in the blanks.

1. Range Olympic Mountains belong to
2. Highest peak in the Olympic Mountains
3. Range Mount McKinley is in
4. Range just north of the Sierra Nevada
5. Highest peak in the 48 states
6. Range in California
7. National Park in Sierra Nevada
8. Highest Mountain in North America
9. National Park in Sierra Nevada
10. Range after Alaska Range
11. National Park in Sierra Nevada

1. P A C I F I C C O A S T
2. M O U N T O L Y M P U S
3. A L A S K A
4. C A S C A D E S
5. M O U N T W H I T N E Y
6. S I E R R A N E V A D A
7. S E Q U O I A
8. M O U N T M C K I N L E Y
9. K I N G S C A N Y O N
10. A L E U T I A N
11. Y O S E M I T E

Death Valley National Monument

Name _____

Death Valley is a desert. The lowest land surface in the Western Hemisphere, 282 feet below sea level, is located in Death Valley. Death Valley's highest point, 11,049 feet, is on top of Telescope Peak in the Panamint Mountains.

Death Valley was named by some pioneers in 1849, who after crossing it, could not see how any life could survive in its dry, hot climate. Usually less than two inches of rain falls in a year. Summer temperatures average 125°F a day. The hottest temperature ever recorded was 135°F. The lowest temperature ever recorded was 15°F. Winter temperatures usually average 90°F. Some animal and plant life have been able to adapt to the desert conditions. In the 1800's, miners lived and worked in the valley. The towns they lived in are ghost towns now. But their burros survived and run wild along with lizards, snakes, desert bighorn sheep and rodents. One man, Walter Scott, lived in Death Valley for thirty years. The "castle" he built is still there.

Write True or False in front of the statements below.

T Some animal and plant life have adapted to Death Valley's climate.
F There are mining towns in Death Valley.
F The highest point in the Western Hemisphere is located in Death Valley.
T Death Valley is a desert.
F Death Valley was named by some miners in 1849.
F Living is easy in Death Valley.
T The highest point in Death Valley is in the Panamint Mountains.
T It is usual for Death Valley to receive less than two inches of rain in a year.
F Visitors to Death Valley will see no signs of life.
F Normal summer temperatures are 135°F.
T The climate in Death Valley is hot and dry.
T Winter temperatures average 90°F.

Badlands National Park

Name _____

Rain, wind and frost have carved ravines, ridges, low hills and cliffs in the South Dakota prairie. These same weather conditions have also exposed the rock so the layers that were laid down millions of years ago are clearly visible. Prehistoric animal and swamp plant fossils are part of the layers, which means the Badlands were once warm and moist. Since the land has been known to white man, it has been dry and bare. The temperatures are very hot in summer and cold in winter. The Badlands were named because they were difficult for the settlers to cross and impossible to farm. Some wildlife has been able to adapt to the Badlands. Unscramble their names below to see what some of the wildlife looks like today in Badlands National Park.

CUYAC YUCCA
3 2 1 5 4

COTTONWOOD
N O W O D O T O T C
6 2 7 5 10 9 4 8 3 1

YTCOOE COYOTE
3 5 1 4 2 6

DREGAB BADGER
3 6 5 4 2 1

GOLDEN EAGLE
G D O N L E   A L E G E
1 4 2 6 3 5   8 10 11 9 7

RATTLESNAKE
S A T E R T K L E N A
7 2 4 11 1 3 10 5 6 8 9

Everglades National Park

Name _____

Everglades National Park is a subtropical region in southern Florida. It is the third largest national park in the United States and one of the country's wettest. The Park includes Ten Thousand Islands on the Gulf of Mexico and the Big Cypress Swamp. The park is a small part of the Florida Everglades. Half of the park is under seawater. Raised islands, called hammocks, are scattered in the marshy sawgrass. The park as a refuge provides protection for many plants, animals and birds.

Discover some of the Everglades wildlife. Label each picture with one of the four titles to the right. Color the finished pictures.

ALLIGATOR AND MANATEE
EGRET
ORCHID
MANGROVE AND CYPRESS TREES

ORCHID

MANGROVE AND CYPRESS TREES

ALLIGATOR AND MANATEE

EGRET

Answer Key

Grand Teton National Park

Name _____

Grand Teton National Park is an area of mountains, lakes and forests in northwestern Wyoming just six miles south of Yellowstone National Park. The two parks are connected by the John D. Rockefeller Memorial Parkway. Grand Teton National Park was established in 1929 after it was unable to become a part of Yellowstone. In 1950 Congress enlarged Grand Teton by making Jackson Hole National Monument a part of it.

The Grand Teton mountains are the youngest of the Rocky Mountain system. About nine million years ago, part of the earth's crust moved upward to form the mountains, and part moved downward to form the valley floor (Jackson Hole). Seven of the Teton peaks rise 12,000 feet above sea level. The Grand Teton is the highest. Several mountain lakes are at the foot of these steep mountains on the valley floor. Jackson Lake is the largest. The Snake River flows out of it.

The names of five other glacial lakes are below. Unscramble them and write their names on the blank lines.

N E N Y J _Jenny_ Lake
3 2 4 5 1

G T A R T G A _Taggart_ Lake
4 1 5 6 7 3 2

P L E H P S _Phelps_ Lake
5 4 3 2 1 6

G L H E I _Leigh_ Lake
4 1 5 2 3

L R E B Y A D _Bradley_ Lake
5 2 6 1 7 3 4

Page 69

Great Smoky Mountains National Park

Name _____

Great Smoky Mountains National Park is on the boundary between North Carolina and Tennessee. It is part of the Appalachian mountain system. Almost all of the Smoky mountains lie within the park. Great Smoky Mountains National Park has sixteen peaks above 6000 feet. Clingman's Dome is the highest. There are over 150 different kinds of trees, and hundreds of shrubs and flowering plants, including several kinds of orchids. Indians called these mountains SHA-GONIGEI which means "little blue smoke". The thick vegetation that covers the park combines with the water to produce a water vapor that covers the park with a smoky mist or haze. That is how the mountain range was named.

Below are some names of some of the mountain peaks over 6000 feet. Unscramble their names and write them on the blanks. The circled letters will spell out the name of the area's first settlers. _C H E R O K E E_

S O L N C I L _C⊙L L I N S_
7 2 3 6 1 5 4

C P M H N A A _C H⊙P M A N_
1 4 5 2 7 6 3

T M E L O T R C E _Mt. L⊙ Co⊙te_
2 1 4 3 6 8 7 5 9

Y O G U T _Guy⊙t_
3 4 1 2 5

P T K E R A H _⊙ephart_
3 7 1 2 6 5 4

G B I E A L E C O H T C A O B O N K
3 1 2 14 5 8 13 4 9 12 6 11 7 10 18 17 16 15

Big Catalooche⊙ Knob

Page 70

Petrified Forest National Park

Name _____

Petrified Forest National Park in Arizona was once a flat floodplain with streams wandering over it. Large trees grew to its south. Reptile-like animals lived in the area too. The trees fell and were washed onto the floodplain. The land sank. The trees and other wildlife were buried under mud, silt and ash. Because no air could reach the trees, they changed from wood to rock—called petrified wood. Millions of years later when the earth moved, it raised the trees above ground. The fallen trees were exposed as petrified wood. Fossils of other ancient plants and animals were also raised. That is how we know today what the area looked like 225 million years ago. When the park was "discovered", souvenir hunters took the wood. In order for it not to disappear, six forests were set aside, and a national park was established.

Work the math problems below. The answers will spell out what scientists discovered recently in the park. Use the code box on the left to decode the message. Write it out under the problems. Read the answer down in columns.

CODE BOX	
A = 13	N = 26
B = 12	O = 25
C = 11	P = 24
D = 10	Q = 23
E = 9	R = 22
F = 8	S = 21
G = 7	T = 20
H = 6	U = 19
I = 5	V = 18
J = 4	W = 17
K = 3	X = 16
L = 2	Y = 15
M = 1	Z = 14

5 + 5 = _10_ D   9 + 12 = _21_ S

22 − 17 = _5_ I   8 − 5 = _3_ K

14 + 12 = _26_ N   18 − 9 = _9_ E

17 + 8 = _25_ O   11 − 9 = _2_ L

28 − 7 = _21_ S   6 + 3 = _9_ E

7 + 6 = _13_ A   12 + 8 = _20_ T

25 − 6 = _19_ U   13 + 12 = _25_ O

17 + 5 = _22_ R   19 + 7 = _26_ N

A _dinosaur skeleton_ nicknamed Gertie

Page 71

Sequoia, Kings Canyon and Redwood National Parks

Name _____

Sequoia, Kings Canyon and Redwood National Parks are in California. Sequoia and Kings Canyon are next to each other on the western slopes of the Sierra Nevada range in the central part of the state. Redwood National Park is along the Pacific coast in northern California and southern Oregon. Sequoia and Kings Canyon are two separate parks, but are run as one by the National Park Service. They contain some of the highest peaks of the Sierra Nevada and some of the oldest and largest trees in the world. Mount Whitney, the tallest mountain in the 48 states, is in Sequoia National Park. General Sherman tree in Sequoia has the largest amount of wood, making it the largest tree in the world—272.4 feet high and 101.6 feet around. Redwood National Park has the world's tallest tree—368 feet tall.

Sequoia Redwood

Write the answers to the questions below on the blanks. When you have answered all the questions, some letters will have a number under them. Write that letter above its number below to learn something about the difference between redwood and sequoia trees.

In what park is the world's largest tree? _S e q u o i a_
                  10 5 8

What is its name? _General Sherman_
                  9   1

In what mountain range are Kings Canyon and Sequoia National Parks located? _S i e r r a N e v a d a_
       7   2   3

In what state are these parks? _C a l i f o r n i a_
               13   12

What is the name of the tallest peak in the 48 states? _Mount Whitney_
            11   4   6

_R e d w o o d t r e e s a r e_
1 2 3 4 5 6 3 6 1 2 2 7 8 1 2

_t a l l e r_
6 8 9 9 2 1

_S e q u o i a s a r e f a t t e r_
7 2 10 11 5 12 8 7 8 1 2 13 6 6 2 1

Page 72

Answer Key

Name _____

Yellowstone National Park

Name _____

Old Faithful

Yellowstone National Park is mainly in the northwest corner of <u>Wyoming</u>, but it also extends into <u>Idaho</u> and <u>Montana</u>. It is famous for many natural wonders, but most of all for its more than 200 <u>geysers</u> and thousands of <u>hot springs</u>. Beginning about 2 million years ago, volcanic eruptions occurred here. What is now the park's central portion collapsed forming a <u>basin</u>. The heat under the earth that caused the ancient eruptions is still responsible for the geysers and hot springs that are in the park today. <u>Old</u> <u>Faithful</u> is the most famous geyser. It erupts boiling water about every 65 minutes. <u>Steamboat</u> <u>Geyser</u> is the world's largest. It spouted 400 feet in the air for a record. <u>Grand</u> <u>Prismatic</u> <u>Spring</u> is the largest spring in Yellowstone. <u>Fountain</u> <u>Paint</u> <u>Pots</u> is a series of hot springs and bubbling pools formed by steam and gases rising from holes in the ground. At <u>Mammoth</u> <u>Hot</u> <u>Springs</u>, flowing water deposits limestone that builds terraces. <u>Norris</u> <u>Geyser</u> <u>Basin</u> has hundreds of geysers and hot spring pools. It is the hottest and most active thermal area in Yellowstone. Just north of Yellowstone Lake, the earth's surface is rising a little less than an inch each year. This suggests there will be future volcanic activity.

Answer the questions below. Let the underlined words above help.

In what state is most of Yellowstone National Park located? <u>Wyoming</u>
Into what other states does it extend? <u>Idaho and Montana</u>
For what is Yellowstone best known? <u>hot springs</u> and <u>geysers</u>
Name the most famous geyser. <u>Old Faithful</u>
What is a series of hot springs? <u>Fountain Paint Pots</u>
Where in the park are hundreds of geysers and hot springs? <u>Norris Geyser</u> <u>Basin</u>
Which spring is the largest? <u>Grand Prismatic Spring</u>
What was formed after the volcanic eruptions? <u>a basin</u>
Which geyser is the largest? <u>Steamboat Geyser</u>
Where are terraces formed? <u>Mammoth Hot Springs</u>

Terraces at Mammoth Hot Springs

Page 73

Name _____

Wildlife In Yellowstone National Park

Yellowstone is the largest wildlife preserve in the United States. Over forty kinds of animals and two hundred birds may be seen there. Look for a bison, elk, grizzly bear, mule deer, trumpeter swan, bighorn sheep, coyote, osprey and pelican among the lodge pole pine in the hidden picture below. Each animal is numbered. Write its name next to its number under the picture.

1. <u>bison</u>   4. <u>mule deer</u>   7. <u>coyote</u>
2. <u>elk</u>   5. <u>trumpeter swan</u> 8. <u>osprey</u>
3. <u>grizzly bear</u>   6. <u>bighorn sheep</u>   9. <u>pelican</u>

Page 74

Yosemite National Park

Name _____

Yosemite National Park is in California in the Sierra Nevada mountains. Five hundred million years ago, it was all under water. Eventually the earth began to twist and turn. Mountains rose out of the sea. The Merced River cut deep into the rock carving a V-shaped valley. But glaciers moved back and forth making it U-shaped. The glaciers cut off many valleys being shaped by smaller streams and caused the formation of "hanging valleys" from which many of Yosemite's waterfalls descend. <u>Yosemite</u> Falls has upper and lower falls. There is a rainbow at the base of <u>Vernal</u> Falls. <u>Nevada</u> Falls was named "Yo-wipe" by the Indians because it means twisted falls in Indian. <u>Bridalveil</u> Falls looks like a bride's headdress. <u>Illilouette</u> Falls is one of the smaller falls descending only 370 feet. <u>Ribbon</u>, <u>Sentinel</u> and <u>Silver Strand</u> Falls usually run only in the spring when there is a lot of moisture. <u>Waterwheel</u> Falls is a series of pinwheels of water in the Hetch Hetchy Valley of the park.

Circle the names of the waterfalls in the puzzle. The names read →↑↓↗↘ Write their names below the puzzle in alphabetical order.

1. <u>Bridalveil</u>   4. <u>Ribbon</u>   7. <u>Vernal</u>
2. <u>Illilouette</u>   5. <u>Sentinel</u>   8. <u>Waterwheel</u>
3. <u>Nevada</u>   6. <u>Silver Strand</u> 9. <u>Yosemite</u>

Page 75

More About Yosemite

Name _____

Yosemite National Park has over 760,000 acres of land with 700 miles of trails. From the Valley Visitor Center in the middle of the park, one can set out to explore many of the park's sights. Follow the directions and mark the map below.

With orange, trace over the road from the east entrance, over Tioga Pass, California's highest automobile pass. Stop at the Tuolumne Meadows Visitor Center and go on to the Valley Visitor Center.

With red, trace over the road from the Valley Visitor Center to Glacier Point where you can get a good view of Yosemite. Circle some of Yosemite's famous landmarks: three rock masses; Cloud's Rest, El Capitan and Half Dome; two waterfalls; Bridalveil and Yosemite; and Mirror Lake.

With blue, trace over the road from the Valley Visitor Center to the Pioneer Yosemite History Center to see some historic buildings and horse drawn carriages.

With green, trace over the road from the museum to Mariposa Grove to see the oldest Sequoia tree, the "Grizzly Giant".

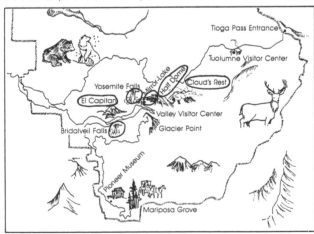

Page 76

Answer Key

Do You Know These Symbols?

Name _____

Written in the box below are names of some symbols of our heritage. Below the box are pictures of these symbols. Write the name for each one on the line under its picture. Color each symbol as you are directed.

> Eagle   Washington Monument   Flag   Liberty Bell
> Statue of Liberty   United States Capitol   The White House
> Mount Rushmore   Jefferson National Expansion Memorial

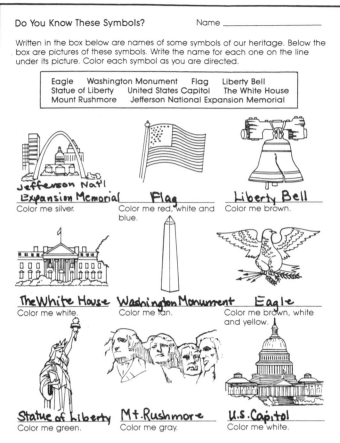

Jefferson Nat'l Expansion Memorial
Color me silver.

Flag
Color me red, white and blue.

Liberty Bell
Color me brown.

The White House
Color me white.

Washington Monument
Color me tan.

Eagle
Color me brown, white and yellow.

Statue of Liberty
Color me green.

Mt. Rushmore
Color me gray.

U.S. Capitol
Color me white.

Page 77

Mount Rushmore

Name _____

Mount Rushmore is a monument of four American President's heads. It is carved into the hard granite Black Hills of South Dakota. The heads stand for the birth and growth of our country. George Washington represents the fight for independence and the start of our nation. Thomas Jefferson stands for the belief in a representative government. Abraham Lincoln represents the unity of the states and equality for all citizens. Theodore Roosevelt stands for the part the United States would take in the world during the twentieth century. The heads were designed by a sculptor named Gutzon Borglum. It took fourteen years to make the monument. Follow the directions below.

Mark a blue X on Jefferson's head. Circle Lincoln with black. Draw an orange line under Washington. Color Roosevelt's face yellow.

Solve the crossword.

DOWN
1. President who believed the government should represent the people
2. Sculptor of Mount Rushmore
3. President who kept the states together
4. Mount Rushmore is in South _____.

ACROSS
5. Number of years it took to build Mount Rushmore
6. President that stood for future of the United States
7. Father of our country

Page 78

The Alamo

Name _____

The Alamo was built as a Catholic mission in 1718. It was not called the Alamo until later. Alamo is a Spanish word for the cottonwood trees that surrounded the mission. The Alamo was used sometimes as a fort. In 1836 it was the scene of a battle between Mexico and 150 Texans who wanted to keep Texas independent. All the Texans were killed, including Davy Crockett. About a month later, behind the battlecry, "Remember the Alamo", General Sam Houston led new forces in a surprise attack on the Mexican troops. He overthrew them. Today the Alamo is a historic structure in the center of San Antonio, Texas.

Find a picture of the Alamo, its flag, Davy Crockett and Sam Houston in the picture below. Color the fort brown, the flag blue and white, Davy Crockett orange and Sam Houston pink.

Page 79

Stone Mountain

Name _____

Stone Mountain is a state park near Atlanta, Georgia, and contains the largest stone mountain in North America. It is called Stone Mountain! Its highest point is 700 feet above the base. It is made of granite. The park has a lake, beach, golf course, museum, a restored plantation and a skylift that carries 50 people at a time to the top of the mountain.

A sculpture is carved into the side of the mountain. It honors the men who fought courageously in the Civil War. Three artists took 46 years to complete it.

There are three southern leaders on horseback on the sculpture. Their names are scrambled below in the order they appear on the sculpture. Unscramble their names and write them on the lines. Color as directed.

N E R F E J F O S   D V S A I
9 2 6 3 5 1 4 8 7   10 12 14 11 13
Jefferson   **Davis**
Color him green.

B E T R R O   E.   E E L
3 4 6 1 5 2   7   10 9 8
Robert   **E.**   **Lee**
Color him red.

T E A S N L O L W   A S J K N C O
2 5 7 1 4 8 3 9 6   11 14 10 13 16 12 15
Stonewall   **Jackson**
Color him blue.

Page 80

Answer Key

Page 81

Alcatraz

Name _____

Alcatraz was a prison for the most dangerous <u>criminals</u>. It was built on a <u>twelve</u> acre rock island, one mile off the coast of <u>California</u>. If a prisoner tried to escape, he <u>drowned</u> or was recaptured before reaching the mainland.

Alcatraz was a prison as described above from 1934 to 1963. Earlier it had served as a <u>barracks</u> for Civil War <u>soldiers</u> and a military prison. In 1969 a group of Indians took it over.

In 1972 it became part of the Golden Gate National Recreation Area. The <u>prison</u> still stands on Alcatraz Island in San Francisco Bay and may be visited by tourists.

Answer the questions below using the underlined words from above. When they are written in the correct blanks, the letters reading down in the boxes will spell what Alcatraz means in Spanish.

What was Alcatraz? — P R I S O N
What might have happened to an escaping prisoner? — D R O W N E D
How many acres are on Alcatraz Island? — T W E L V E
Who lived there during the Civil War? — S O L D I E R S
What state is one mile from Alcatraz? — C A L I F O R N I A
In what did the soldiers live during the Civil War? — B A R R A C K S
What sort of people lived there from 1934–1963? — C R I M I N A L S

What does Alcatraz mean in Spanish? pelican

Alcatraz has a nickname. Unscramble the letters below. Write its nickname on the lines following.

H E T   C O R K
2 3 1   6 5 4 7

T H E   R O C K
1 2 3   4 5 6 7

Page 81

Page 82

Ellis Island

Name _____

Ellis Island is in New York Harbor. It was the place most people coming to the United States for the first time had to go before being allowed to stay in America. They were examined there by doctors and checked to be sure they had no criminal record. If they were found to have a disease or to have something wrong with their past, they were sent back to the country from which they had come.

Ellis Island is no longer used as an immigration station, but it stands as a symbol of the time when hundreds of foreigners passed through immigration. Thirty or more buildings remain on Ellis Island near ruin. They are being restored and should be open for the public to visit sometime after 1990.

America is a melting pot. Originally Americans came from other countries besides the United States. The names of many countries from which the immigrants came are scrambled below. Unscramble them to learn the names of some of these countries.

A Y I L T — Italy
3 5 1 4 2

P N A O D L — Poland
1 5 4 2 6 3

M A N R O I A — Romania
3 4 5 1 2 6 7

G R U N A H Y — Hungary
4 6 2 3 5 1 7

L I E A R D N — Ireland
4 1 3 5 2 7 6

N E F A R C — France
4 6 1 3 2 5

S I R S U A — Russia
3 5 1 4 2 6

Y N G R E M A — Germany
7 6 1 3 2 4 5

G L E D N A N — England
3 4 1 7 6 5 2

O C L H Z S K O C V A A E I — Czechoslovakia
9 4 8 5 2 7 12 6 1 10 14 11 3 13

Page 82

Page 83

Appomattox Court House

Name _____

Appomattox Court House was a small village of about 150 people in Virginia. It had a few homes, stores, lawyers' offices, a tavern and a courthouse. It was the county seat for Appomattox County, a farming area. On April 9, 1865, General Robert E. Lee, commander of the North Virginia Army, met with General Ulysses S. Grant, commander-in-chief of the Union army, in Wilmer McLean's farmhouse. Lee surrendered his men to Grant. Grant only asked that the Confederate men not take up arms again against the United States. He freed them to return to their homes. When news of Lee's surrender reached the Confederate armies still fighting out in the fields, the Confederate soldiers put down their weapons there and went back to their homes, too. The Civil War was over. Today Appomattox Court House is a national historical park open to the public.

Write either TRUE or FALSE in front of each statement about Appomattox Court House.

True Appomattox Court House was a small village.
True It is located in Virginia.
False Grant surrendered to Lee.
True Lee was commander of the North Virginia Army.
True The Union won the Civil War.
False Lee and Grant met in an old estate.
True The courthouse was the county seat.
True Grant was the commander-in-chief of the Union armies.
False Lee surrendered to Grant in August.
False After the Civil War, the Confederate soldiers were put in jail.
True Today Appomattox Court House is open to the public.
True Most of the people in Appomattox County were farmers.
True Wilmer McLean had a farmhouse.
False When Lee surrendered, all fighting stopped immediately.
True There were a few homes and offices in Appomattox Court House.

Page 83

Page 84

Independence Hall

Name _____

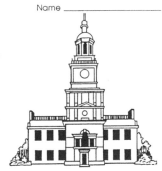

Independence Hall was not always called by that name. It was built in Philadelphia as the state house for Pennsylvania and was called the Old State House. Its purpose was for Pennsylvania's lawmakers to meet there.

In 1775 America's early statesmen used the Old State House when they met to write the Declaration of Independence. Eleven years later the Constitution of the United States was written there. The Liberty Bell once hung in its tower. In 1818 Philadelphia bought the Old State House. In 1824 General Lafayette renamed it Independence Hall.

Today it is a National Historical Park. Visitors from all over the world come to see the "Shrine of Independence" where the United States government got started.

Circle every third letter below to learn one thing visitors will see when they visit Independence Hall. Write it out on the lines at the bottom of the page.

EGTIEHMQE   INSXUIBW LAOVZDEMCR   PUILKNJHKGFSDATQWAERNTYD

MZUXCSVBENMD   GMBPTY   WRTPUHIEE   QRSPUIVXGAINVRESDRABS

PIODAF   VIBPPOSRTVGH   VYTWIHQPE

IEDGHEPACXYLMBAEDRZYAOGTUNIVFOMSN   PWOVHF

OPINMNPADIEEBYPWSEXBNPPDAIECWNIQCQUE   ORAXFNOID   FGTBUHMAE

BACKEOLDNRESUPTRAINOTRQUYATPAIMBOANN

The silver inkstand used by
the signers of both the
Declaration of Independence
and the Constitution

Page 84

Answer Key

Presidential Homes
Name _____

Read about two presidential homes and connect the dots to see how each one looks.

Mount Vernon was not always called Mount Vernon. George Washington's father built the main house in the 1730's and called the plantation "Little Huntington Creek". Mount Vernon was built in stages from 1742 to 1787. Washington's half brother, Lawrence, named it Mount Vernon when he headed the estate. When Lawrence died, Mount Vernon became Washington's. By the time Washington became President, the estate had five working farms and several outbuildings.

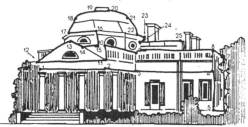

When Thomas Jefferson was fourteen years old, his father died and left him 2,750 acres in Virginia. Nine years later, in 1768, Jefferson designed and began to build Monticello on a hilltop. In Italian Monticello means "little mountain". The house was not completed until 1809. It was filled with many of Jefferson's inventions, such as a dumb waiter, calendar clock and a revolving desk. As in many homes like Monticello and Mount Vernon, the kitchens were not in the main structure. Do you know why?

Page 85

The Statue of Liberty
Name _____

The Statue of Liberty is a huge national monument that stands on Liberty Island in New York Harbor. It was given to the United States by France on July 4, 1884, as a symbol of friendship and liberty. France paid for the statue. The American people paid for the pedestal on which it stands.

The statue was designed and made in France. It was shipped to the United States in 214 boxes. The statue was put together and placed on its pedestal which had been built over old Fort Wood. The statue was completed in 1886.

The statue became a national monument in 1924. When the statue was 100 years old, there was a big birthday party on July 4, 1986.

Color the statue green.
Color the pedestal tan.

Unscramble the letters after each question below to find the answer. Write it on the line following the scrambled word.

What was the name of the fort upon which the Statue of Liberty stands?
ORFT ODOW _Fort Wood_

Who gave the statue to the United States? CNREFA _France_

On what island does the statue stand? BIRLYET _Liberty_

In what Harbor is Liberty Island? WNE ROYK _New York_

What country paid for the pedestal? RMAEACI _America_

Page 86

The Liberty Bell
Name _____

The first Liberty Bell was ordered from London in 1752 by the Pennsylvania legislature in observance of their fifty years of liberty as a colony. The first time the bell was rung, it cracked. A new bell was recast by a Philadelphia foundry using the same metal. The second bell did not have a good tone, so a third bell was made.

The bell was the official town bell. It rang to call town meetings, announce the end of wars and deaths of great men. It also served as a fire alarm. It rang to announce the signing of the Declaration of Independence in 1776.

During the Revolutionary War, the Liberty Bell was moved to Allentown, Pennsylvania, so the enemy could not be able to seize it and melt it down for ammunition. In 1778 when the British had gone from Philadelphia, the bell was returned to its place in the State House and rang again. It rang every Fourth of July until 1835. Not until the 1840's was it called the Liberty Bell.

Work the math problems below to find what the other names were for the bell. Then look to see what letter goes with each answer. Write the letter below each answer.

A=1	B=2	C=3	D=4	E=5	F=6	G=7	H=8	I=9
J=10	K=11	L=12	M=13	N=14	O=15	P=16	Q=17	
R=18	S=19	T=20	U=21	V=22	W=23	X=24	Y=25	Z=26

```
  8    16   10      4    9    2   20    8   13   15     7    6    9   12
 +7   -4   -6     +5   +5  +15  -8   -1   -3   +3    +8   +6   -6   -7
 ───  ───  ───    ───  ───  ─── ───  ───  ───  ───   ───  ───  ───  ───
 15   12    4      9   14    5   12    5   14    5     15   14    3   15
  O    L    D       I   N    D    E    N    D    E     O    N    C    E,
```
O L D   I N D E P E N D E N C E,

```
 19    4    9      10   11    6   18   11      3   13   14   20    1      8   14    4   16
 -4   +8   -5     -4   +9   +9  -5   +7    -1   +4  -7   -1   +4   -6   -9   +8   -4
 ───  ───  ───    ───  ───  ─── ───  ───   ───  ───  ─── ───  ───   ───  ─── ───  ───
 15   12    4      19   20    1   20    5     8   15   21   19    5     25   12    2
  O    L    D       S    T    A    T    E      H    O    U    S    E     B    E    L    L
```
O L D   S T A T E   H O U S E   B E L L   OR

```
 11    0   15    6      9   10      10   12    9
 -9   +5   -3   +6    +6   -4    +10   -4   -4
 ───  ───  ───  ───   ───  ───   ───   ───  ───
  2   15   12   12    15    6     20    8    5
  B    O    L    L     O    F      T    H    E
```
B E L L   O F   T H E

```
  9   11   11   20    7   10   15   12   17    7
 +9   -6  +11   -5   +5  +11   +5   -3   -2   +7
 ───  ───  ───  ───  ───  ───  ───  ───  ───  ───
 18    5   22   15   12   21   20    9   15   14
  R    E    V    O    L    U    T    I    O    N
```
R E V O L U T I O N

Page 87

The Star-Spangled Banner
Name _____

The Star-Spangled Banner was written by Francis Scott Key. Key had permission from President Madison to meet with the British to try and have a friend that had been captured freed. Key sailed into Baltimore's harbor to meet the British Admiral who agreed to set Key and his friend free after the attack on Baltimore.

Key watched the flag above Fort McHenry all through the day, September 13, 1814. That night he listened to bombs bursting in the air. At dawn on September 14th, when he saw the flag still flying over Fort McHenry, he knew the British had lost and that Baltimore was still a city. He was so happy that he wrote a poem to express how he felt. The poem was set to music written by an Englishman for a different poem.

Fill in the missing words below. They are underlined above. The circled letters in your answers will spell out the first five words of the Star-Spangled Banner.

M a d i s (o) n was the President when the Star-Spangled Banner was written.
Francis Scott Key went to meet
the B r i t i s h   A d m i r a l.
The battle began during the d a (y).
The flag above Fort M (c) H e n r y lasted through the battle.
At d (a) w (n) it was still there.
K e (y) was so happy that he w r o t e a poem.
It was set to m (u) (s) i (c) written by
an E n g l i s h m a n for a different p o (e) m.
Write the first five words of the Star-Spangled Banner.
O say can you see

Page 88

Our 50 States IF8749   124   © 1992 Instructional Fair, Inc.

Answer Key

Uncle Sam

Name _____

Uncle Sam is not a real person. He is a symbol of the United States. He often appears on posters asking the American people to do something good for their country. Although he has been around for over one hundred fifty years, Congress did not recognize him as a national symbol until 1961.

First draw Uncle Sam below by connecting the dots. Then color him by using the color code.

red = A
white = B
blue = C
gray = D
yellow = E

Page 89

The Flag of the United States

Name _____

The Continental Congress left no record of why they chose red, white and blue as the colors of the flag. The stripes in the flag were to stand for the original colonies. In 1777 Congress said the flag should have thirteen stars, but it did not say how they should be arranged. Every time a state was added to the Union, another star was added to the flag. The stars had no special arrangement until 1912. Since then the President has ordered how the stars are to be arranged.

Connect the dots in the boxes below. Color each flag.

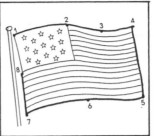

The Flag from 1776-1777

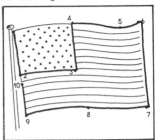

The Flag Now

How are they alike? **Same number of stripes**
How are they different? **Number of stars**
Over the years the United States flag has had many names. Unscramble the letters below and write them in order on the lines following to learn some of the flag's other names.

L O D   L Y O R G   **OLD GLORY**
2 1 3   5 8 6 7 4

A R S T S   A D N   T I E R P S S   **STARS AND STRIPES**
3 4 1 2 5   6 8 7   10 12 14 11 13 9 15

T A R S N G S P E A L D   N E R B N A   **STAR-SPANGLED BANNER**
2 3 4 1 6 5 11 7 10 12   15 17 13 16 14

Page 90

The Pledge of Allegiance

Name _____

The Pledge of Allegiance is a promise to be true to the United States. It was first recited in 1892 by children saluting the flag. President Benjamin Harrison had asked there be patriotic celebrations in America's schools to mark the 400th anniversary of the country's discovery. Francis Bellamy wrote the original pledge. It was changed two or three times before it had the wording we use today.

Cross out every third letter below beginning with the third one. The letters that remain will spell out the Pledge of Allegiance. The lines (—) between the letters are to help divide the words. Write the pledge on the lines at the bottom of the page.

I—PALEXDGME—AFLLUEGRIARNCNE—THO—TIHEN—FLSAGT—OFE—THAE—
UHNIBTERD—SCTAPTEOS—OJF—AKMEGRIDCAS—ANOD—TRO—TSHEW—
RESPUZBLAICY—FOER—WEHIRCHS—ITS—STRANLDSI—ONCE—NEATRIOUN—
UENDLERS—GOLD—IENDRIVAISSIBLLES—WIATHY—LIWBEVRTEY—ALNDE—
JUXSTRICEE—FLORT—ALIL

I pledge allegiance to the flag
of the United States of America
and to the Republic for which it
stands, one nation under God
indivisible with liberty and
justice for all.

Page 91

The Declaration of Independence

Name _____

In CONGRESS July 4, 1776

The Declaration of Independence is a document written in 1776 in which America's thirteen colonies declared their freedom from British rule. England had been deciding many of the laws for the new country and had been taxing the colonists too much.

The colonists were very unhappy. The Declaration of Independence lists their complaints against the King of England and gives good reason for the breaking away of the colonies from England.

The Second Continental Congress appointed a committee to write the declaration. Thomas Jefferson was the main author of the document. Others on the committee were Robert Livingston, John Adams, Roger Sherman and Benjamin Franklin. The Declaration of Independence is signed by fifty-six delegates. John Hancock signed it First on July 4, 1776, because he was President of the Second Continental Congress.

Fill in the blanks below with the underlined words from above. Write out the circled letters in order in the box at the bottom of the page. They will spell the name of the holiday on which we celebrate the birth of our country.

1. **Thomas Je(f)ferson** was the main author of the Declaration of Independence.
2. The **Sec(o)nd** Continental Congress appointed a committee to write the Declaration of Independence.
3. The colonists broke away from British **r(u)le**.
4. **R(o)bert Livingston**, **Roger Sherma(n)** and **J(o)hn Adams** were other authors of the Declaration of Independence.
5. The **(F)irst** to sign the Declaration of Independence was John Hancock.
6. The fifth author of the Declaration of Independence was **Ben(j)amin Franklin**.
7. The Declaration of Independence is a **doc(u)ment** of freedom.
8. States were called **col(o)nies** in 1776.
9. **Fift(y)-six** delegates signed the Declaration of Independence.

Fourth of July

Page 92

Answer Key

The Constitution of the United States

Name _____

The Constitutional Convention met in Philadelphia in 1787. Fifty-five delegates from twelve of the thirteen states were there. George Washington was President of the Constitutional Convention. The Constitution was written between May 25 and September 17, 1787. Only thirty-nine of the delegates signed it. William Jackson, secretary to the convention, signed as witness to the signatures. After it was signed, copies were made and sent to all the states for their approval. The Constitution was ratified first by Delaware on December 7, 1787.

The United States became a united nation with the signing of the Constitution. The Constitution's seven articles and twenty-six amendments are the basis of our laws. The Constitution lists the three branches of government—executive, legislative and judicial. It lists the rights of the people. Today the Constitution is on display, under glass, in Washington, D.C. in the National Archives Building.

Answer TRUE (T) or FALSE (F) to the following statements.

F All states attended the Constitutional Convention.
F George Washington was President of the U.S. in 1787.
T The Constitutional Convention met in Philadelphia.
F Forty delegates signed the Constitution.
T The Constitution established the country's laws.
T The original copy of the Constitution is in the National Archives Building.
F All the delegates signed the Constitution.
T The states had to approve the Constitution.
F The Constitution was written in three days.
T On June 21, 1788, the Constitution was approved.
T The Constitution has seven articles and twenty-six amendments.
T William Jackson was the convention's secretary.
T The formation of the executive, legislative and judicial branches are outlined in the Constitution.
F The original Constitution is in Philadelphia now.
T Several copies of the Constitution were made by the Constitutional Convention.

Page 93

Bill of Rights

Name _____

James Madison presented twelve amendments to the first Congress in 1789. Congress passed ten of them. They are called the Bill of Rights. They became law December 15, 1791. In 1940, President Roosevelt declared December 15th Bill of Rights day.

The Bill of Rights describes freedoms every American has. Some of the rights Americans have are freedom of speech, press, religion and assembly. Every American may say and write what he or she wants, and pray and meet where he or she wants. The Bill of Rights guarantees freedom for all Americans—something that we perhaps just take for granted.

Use the underlined words above to fill in the crossword puzzle.

Across
2. Last name of man responsible for Bill of Rights
6. One of the rights
8. One of the rights
9. One of the rights

Down
1. Number of amendments that are part of the Bill of Rights
3. Month they became law
4. What every American is guaranteed with the Bill of Rights
5. First name of man responsible for the Bill of Rights
7. One of the rights

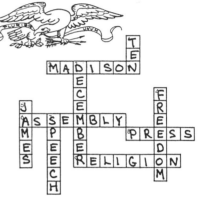

Page 94

The Great Seal of the United States

Name _____

The United States government adopted the Seal of the United States on June 20, 1782. It used the seal on important documents. The American eagle, with a shield on its chest, represents self-reliance. The stripes on the shield come from the flag of 1777. The blue at the top of the shield symbolizes all branches of government. The eagle holds thirteen olive branches in its right talon and thirteen arrows in its left showing its desire to live in peace, but with the capability of waging war. The yellow banner in the eagle's beak has Latin words on it meaning "One (nation) out of many (states)." Above the eagle's head is a circle with thirteen stars inside a golden circle breaking through a cloud.

Color the Great Seal of the United States using the color key to the left.

1 – blue   5 – yellow
2 – brown   6 – white
3 – green   7 – grey
4 – red   8 – black

This is the face of the seal. There is a reverse side. It has never been used as a seal, but it can be found on the back of a dollar bill. Circle the reverse side of the seal on the left.

Page 95

The White House

Name _____

The official residence of the President of the United States has not always been called the White House. It was first called the President's House, then the Executive Mansion. Finally in 1901, Theodore Roosevelt gave it its name, the White House.

One and a half million visitors are given guided tours of the White House's reception rooms every year. They are not allowed above the first floor. The President and his family live on the second floor, and the third floor has guest rooms and servants' quarters. Offices have been added over the years to the East and West wings of the White House.

The answer to each question below is spelled out in the circle next to the question. Begin with the arrow. Go around to the right to find the answer to each question. Write it on the line next to the circle.

1. Who was the first President to live in the White House? John Adams

2. Which President had to leave the White House when the British set fire to it? James Madison

3. In what city is the White House? Washington, D.C.

4. What is the name of one of the rooms visitors may see? State Dining Room

5. On what street is the White House? Pennsylvania

Who lives in the White House now? Ronald Reagan

Page 96

Answer Key

The United States Capitol

Name _____

The United States Capitol is a symbol of the nation and a government office building where Congress meets to make the nation's laws. The building is made of marble. The dome is made of iron and painted white to look like the rest of the Capitol. The Capitol was built in 1800, burned in 1814 and rebuilt. In 1863 the Capitol received a new dome. A statue of a woman named Freedom was placed on top. The distance from the top of the statue to the ground is about 300 feet.

Connect the dots to make a picture of the Capitol.

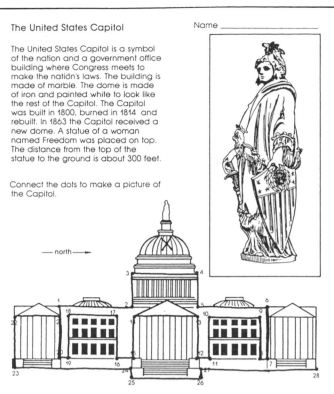

◄— north —►

The Senate meets in the north wing of the Capitol. Circle it. The House of Representatives chamber is in the south wing. Draw a line under it.

The center part of the Capitol is called the Great Rotunda. Mark it with an X.

Page 97

The Supreme Court

Name _____

When the Constitution was written, it provided for three branches of government—the Executive, Legislative and Judicial. The Supreme Court heads the Judicial branch. It is the highest court in the land. It is responsible for carrying out the laws as written in the Constitution.

Congress determines the number of judges, or justices, that sit on the Supreme Court. Since 1869 there have always been nine. The justices are appointed by the President and approved by the Senate. In this way no one branch has all the power. Once a justice is appointed, he or she may hold the position for life. The Supreme Court meets from October to June every year. For 135 years it met in the Capitol. Since 1935 it meets in its own building.

There is an inscription over the entrance to the Supreme Court building which states the Court's belief and duty. To learn what it says, find the answer to each problem. Look to the left to see what letter goes with each answer. Write the letter under the number that represents it.

A = 4
B = 3
C = 2
D = 1
E = 8
F = 7
G = 6
H = 5
I = 12
J = 11
K = 10
L = 9
M = 16
N = 15
O = 14
P = 13
Q = 20
R = 19
S = 18
T = 17
U = 26
V = 25
W = 24
X = 22
Y = 23
Z = 21

$$\begin{array}{c}3\\+5\\\hline 8\\E\end{array}\quad\begin{array}{c}14\\+6\\\hline 20\\Q\end{array}\quad\begin{array}{c}13\\+13\\\hline 26\\U\end{array}\quad\begin{array}{c}9\\-5\\\hline 4\\A\end{array}\quad\begin{array}{c}17\\-8\\\hline 9\\L\end{array}$$

$$\begin{array}{c}7\\+4\\\hline 11\\J\end{array}\quad\begin{array}{c}30\\-4\\\hline 26\\U\end{array}\quad\begin{array}{c}23\\-5\\\hline 18\\S\end{array}\quad\begin{array}{c}9\\+8\\\hline 17\\T\end{array}\quad\begin{array}{c}4\\+8\\\hline 12\\I\end{array}\quad\begin{array}{c}9\\-7\\\hline 2\\C\end{array}\quad\begin{array}{c}14\\-6\\\hline 8\\E\end{array}$$

$$\begin{array}{c}17\\+9\\\hline 26\\U\end{array}\quad\begin{array}{c}7\\+8\\\hline 15\\N\end{array}\quad\begin{array}{c}1\\+0\\\hline 1\\D\end{array}\quad\begin{array}{c}19\\-11\\\hline 8\\E\end{array}\quad\begin{array}{c}25\\-6\\\hline 19\\R\end{array}\quad\begin{array}{c}5\\+4\\\hline 9\\L\end{array}\quad\begin{array}{c}13\\-9\\\hline 4\\A\end{array}\quad\begin{array}{c}12\\+12\\\hline 24\\W\end{array}$$

The inscription over the entrance to the Supreme Court building reads Equal justice under law

Page 98

The Gettysburg Address

Name _____

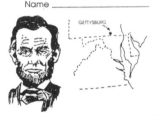

On November 19, 1863, four months after the bloodiest battle of the Civil War, Lincoln delivered his famous Gettysburg Address on the site of the battle. The purpose of his speech was to dedicate a part of the battlefield as a cemetery. Then the over 7,000 soliders from both the North and South who had lost their lives could be buried. He worked hard to make the speech perfect. He chose his words carefully because the Civil War was still being fought. He wrote five different versions of it. Only the last one is signed. A copy of it is carved in stone inside the Lincoln Memorial.

Four score and seven years ago our fathers brought forth, upon this continent, a new nat concentrated in liberty, and

Cross out every other letter below beginning with the second one. The letters that remain will spell out how many words were in the Gettysburg Address.

TWWEOX   HRUNNODTRGEIDT   STERVOENNUTVYE — THWIOL

There were two hundred seventy-two words in the Gettysburg Address.

Do it again to learn how many minutes it lasted.

TPWIOG   MUIGNWUTTHEDSC

The speech lasted two minutes.

Do it one more time. The letters that remain will spell the first six words of the Gettysburg Address.

FPOEUVRX   SPCQOTRAEN   AVNQDE   SWECVLEBNY   YAEAAFRGSM   ARGUOL

The first six words of the Gettysburg Address are:

Four score and seven years ago

One score = 20. Four score = 80. 80 + 7 = 87 How many years was he talking about? 87

Page 99

The Vietnam Veterans' Memorial

Name _____

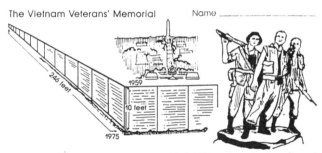

The Vietnam Veterans' Memorial honors all men and women of the United States armed forces who served in the Vietnam War from 1959 to 1975. Congress gave the land for it, but the memorial was built with contributions from business and civic groups and more than 275,000 private citizens. It cost $7,000,000. The memorial was dedicated in 1982. A sixty-foot flagpole and a three-figured sculpture by Frederick Hart were added to the memorial site in 1984.

The memorial is on the Mall in Washington, D.C. set between the Washington Monument and the Lincoln Memorial. The names of the 58,022 men and women who lost their lives or remain missing are carved in its two black granite walls.

Fill in the cross word with the correct numbers.

ACROSS
2. Year Vietnam War began
4. Year sculpture and flagpole were added
5. Number of citizens who helped build memorial with dollars
7. Height of flagpole in feet
9. Number of names carved in walls

DOWN
1. Year Vietnam War ended
3. Length of each wall in feet
4. Height of memorial where two walls join in feet
6. The millions of dollars it cost
8. Year memorial was dedicated

Page 100

Arlington National Cemetery

Name _____

Arlington National Cemetery is in Virginia just across the Potomac River from Washington, D.C. There are over 200,000 Americans buried there. It is run by the United States Army. John F. Kennedy, the 35th President of the United States, Joe Louis, a world heavyweight champion, and Adm. Robert Peary, discoverer of the North Pole, are some of the better known to be buried there along with thousands of military personnel.

Besides the thousands of grave markers, there are several special monuments in the cemetery. To learn what some of them are, find the answer to each math problem. Look to the left to see what letter goes with each answer. Write the letter under the number that represents it. Under that will be a fact about the monument.

A = 2
B = 1
C = 4
D = 3
E = 6
F = 5
G = 8
H = 7
I = 10
J = 9
K = 12
L = 11
M = 14
N = 13
O = 16
P = 15
Q = 18
R = 17
S = 20
T = 19
U = 22
V = 21
W = 24

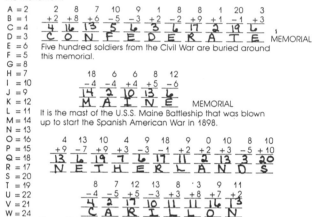

2	8	7	10	9	1	8	8	1	20	3
+2	+8	+6	−5	−3	+2	−2	+9	+1	−1	+3
4	16	13	5	6	3	6	17	2	19	6
C	O	N	F	E	D	E	R	A	T	E

MEMORIAL

Five hundred soldiers from the Civil War are buried around this memorial.

18	6	6	8	12
−4	−4	+4	+5	−6
14	2	10	13	6
M	A	I	N	E

MEMORIAL

It is the mast of the U.S.S. Maine Battleship that was blown up to start the Spanish American War in 1898.

4	13	10	4	9	18	9	0	10	8	10
+9	−7	+9	+3	−3	−1	+2	+2	+3	−5	+10
13	6	19	7	6	17	11	2	13	3	20
N	E	T	H	E	R	L	A	N	D	S

8	7	12	13	8	3	9	11
−4	−5	+5	−3	+3	+8	+7	+2
4	2	17	10	11	11	16	13
C	A	R	I	L	L	O	N

These 49 bells were a gift from the people of the Netherlands.

Page 101

The Tomb of the Unknowns

Name _____

The Tomb of the Unknowns in Arlington National Cemetery was originally called The Tomb of the Unknown Soldier. An unknown American soldier from World War I was laid to rest there on Armistice Day, November 11, 1921. In 1958, the name of the memorial was changed to The Tomb of the Unknowns when an unidentified serviceman from World War II and the Korean War were buried there. On Memorial Day, May 28, 1984, a fourth unknown soldier from the Vietnam War was laid to rest between the World War II and Korean heroes.

The Tomb is guarded 24 hours a day, 365 days a year. The guards belong to the Honor Guard Company, Third U.S. Infantry. The guards have a ritual based on the number 21. It is symbolic of a 21 gun salute—the highest salute given. The guard faces the tomb for 21 seconds, turns and pauses for 21 seconds, and then walks 21 steps, faces the tomb and repeats going in the opposite direction. Back and forth he goes until the guard changes—every half hour in the summer; every hour in the winter. Over 3,000,000 visitors enjoy this ceremony annually.

There are some words written on the tomb. Cross out every other letter below starting with the second one. Write the remaining letters on the blanks to learn what is written on the tomb.

HSEARBEL   REEASRTESC   IUNT   HEOLNIOTRTELDE   GBLEOARSYT
ASNB   AUMTETRHIECYACNA   SNOBLEDVIEERRY   KVNEORWYNM
BEUATN   TAON   GDOSDN

"HERE RESTS IN HONORED GLORY AN AMERICAN SOLDIER KNOWN BUT TO GOD."

Page 102

About the book . . .

The activities in the first half of this book go beyond just the naming of the states and their capitals; the activities provide an understanding of each state — historically, economically and geographically.

The activities in the second half of the book provide interesting facts and colorful descriptions about several of our country's natural attractions, as well as historical and current information about many of our nation's most important symbols.

About the authors . . .

Lee Quackenbush is a professional developer of educational materials for children. After earning her Master's Degree from Memphis State University, she taught remedial reading and all elementary levels for thirteen years.

Claire Norman is an experienced author and veteran teacher, with a Master's Degree in Reading and an Advanced Graduate Certificate from Washington University. Her 33 years of teaching experience include all of the elementary grades, remedial reading, the teaching of art, and directing a Media Center.

Credits . . .

Authors: Lee Quackenbush and Claire Norman
Editor: Lee Quackenbush
Artists: Pat Biggs and Ann Stein
Production: Pat Geasler
Cover Photo: Frank Pieroni
Cover Design: Jan Vonk